T0140018

Smart Innovation, Systems and Technologies

Volume 108

Series editors

The Smart Innovation, Systems and Technologies book series encompasses the topics of knowledge, intelligence, innovation and sustainability. The aim of the series is to make available a platform for the publication of books on all aspects of single and multi-disciplinary research on these themes in order to make the latest results available in a readily-accessible form. Volumes on interdisciplinary research combining two or more of these areas is particularly sought.

The series covers systems and paradigms that employ knowledge and intelligence in a broad sense. Its scope is systems having embedded knowledge and intelligence, which may be applied to the solution of world problems in industry, the environment and the community. It also focusses on the knowledge-transfer methodologies and innovation strategies employed to make this happen effectively. The combination of intelligent systems tools and a broad range of applications introduces a need for a synergy of disciplines from science, technology, business and the humanities. The series will include conference proceedings, edited collections, monographs, handbooks, reference books, and other relevant types of book in areas of science and technology where smart systems and technologies can offer innovative solutions.

High quality content is an essential feature for all book proposals accepted for the series. It is expected that editors of all accepted volumes will ensure that contributions are subjected to an appropriate level of reviewing process and adhere to KES quality principles.

More information about this series at http://www.springer.com/series/8767

Maria Virvou · Fumihiro Kumeno
Konstantinos Oikonomou
Editors

Knowledge-Based Software Engineering: 2018

Proceedings of the 12th Joint Conference on Knowledge-Based Software Engineering (JCKBSE 2018) Corfu, Greece

Springer

Editors
Maria Virvou
Department of Informatics
University of Piraeus
Piraeus, Greece

Konstantinos Oikonomou
Department of Informatics
Ionian University
Corfu, Greece

Fumihiro Kumeno
Nippon Institute of Technology
Miyashiro, Japan

ISSN 2190-3018 ISSN 2190-3026 (electronic)
Smart Innovation, Systems and Technologies
ISBN 978-3-030-07388-6 ISBN 978-3-319-97679-2 (eBook)
https://doi.org/10.1007/978-3-319-97679-2

Softcover re-print of the Hardcover 1st edition 2019

This Springer imprint is published by the registered company Springer Nature Switzerland AG
The registered company address is: Gewerbestrasse 11, 6330 Cham, Switzerland

Preface

This volume summarizes the works and new research results presented at the **12th Joint Conference on Knowledge-based Software Engineering (JCKBSE 2018)**, which took place in August 27–30, 2018, on the island of Corfu, Greece. JCKBSE is a well-established international biennial conference that focuses on the applications of Artificial Intelligence on Software Engineering. The 12th Joint Conference on Knowledge-based Software Engineering (JCKBSE 2018) was organized by the Department of Informatics of the University of Piraeus, the Department of Computer and Information Engineering of Nippon Institute of Technology, and the Department of Informatics of the Ionian University. It was the third time that a JCKBSE took place in Greece.

This year, pretty much like every year, the majority of submissions originated from Japan, while Greece was second. The submitted papers were rigorously reviewed. Finally, 26 papers were accepted for presentation at JCKBSE 2018 and inclusion in its proceedings. The papers accepted for presentation in JCKBSE 2018 address the following topics:

- Architecture of knowledge-based systems, intelligent agents, and softbots
- Architectures for knowledge-based shells
- Automating software design and synthesis
- Decision support methods for software engineering
- Development of multi-modal interfaces
- Development of user models
- Development processes for knowledge-based applications
- Empirical/evaluation studies for knowledge-based applications
- Intelligent user interfaces and human–machine interaction
- Internet-based interactive applications
- Knowledge acquisition

- Knowledge engineering for process management and project management
- Knowledge management for business processes, workflows, and enterprise modeling
- Knowledge technologies for Semantic Web
- Knowledge technologies for service-oriented systems, Internet of services, and Internet of things
- Knowledge technologies for Web services
- Knowledge-based methods and tools for software engineering education
- Knowledge-based methods and tools for testing, verification and validation, maintenance, and evolution
- Knowledge-based methods for software metrics
- Knowledge-based requirements engineering, domain analysis, and modeling
- Methodology and tools for knowledge discovery and data mining
- Ontologies and patterns in UML modeling
- Ontology engineering
- Program understanding, programming knowledge, modeling programs, and programmers
- Software engineering methods for intelligent tutoring systems
- Software life cycle of intelligent interactive systems
- Software tools assisting the development

In addition to technical paper presenters, in JCKBSE 2018, we had the following distinguished researchers as keynote speakers:

1. Prof.-Dr. Demetra Evangelou, Democritus University of Thrace, Greece
2. Prof.-Dr. Fumihiro Kumeno, Nippon Institute of Technology, Japan
3. Prof.-Dr. George A. Tsihrintzis, University of Piraeus, Greece
4. Prof.-Dr. Lefteri Tsoukalas, University of Thessaly, Greece; Purdue University, USA
5. Prof.-Dr. Maria Virvou, University of Piraeus, Greece

We would like to thank **Prof.-Dr. Lakhmi C. Jain** both for acting as **Honorary Chair of JCKBSE 2018** and for agreeing to publish the JCKBSE 2018 proceedings in the form of a volume in the *Smart Innovation, Systems and Technologies* series of Springer, which he edits.

We also would like to thank the authors for choosing JCKBSE 2018 as the forum for presenting the results of their research. Additionally, we would like to thank the reviewers for taking the time to review the submitted papers rigorously. For putting together the Web site of JCKBSE 2018, we would like to thank Mr. Ari Sako. For managing the conference administration system and coordinating

JCKBSE 2018, we would like to thank **Easy Conferences Ltd., Nicosia, Cyprus**. Finally, we would like to thank the **Springer personnel** for their wonderful job in producing this proceedings.

George A. Tsihrintzis
The JCKBSE 2018 General Chairs
Andreas Floros
The JCKBSE 2018 General Chairs
Fumihiro Kumeno
The JCKBSE 2018 General Chairs
Maria Virvou
The JCKBSE 2018 Program Chairs
Fumihiro Kumeno
The JCKBSE 2018 Program Chairs
Konstantinos Oikonomou
The JCKBSE 2018 Program Chairs

Sponsoring Institutions

University of Piraeus—Graduate Program of Studies in Advanced Computing and Informatics Systems

University of Piraeus—Graduate Program of Studies in Informatics

Contents

Ripple Effect Analysis Method of Data Flow Diagrams in Modifying Data Flow Requirements

Jo Heayyoung, Takayuki Omori, and Atsushi Ohnishi(✉)

Department of Computer Science, Ritsumeikan University, Kusatsu, Japan
ohnishi@cs.ritsumei.ac.jp

Abstract. Ripple effects may occur in modifying software requirements. Without considering such ripple effects, errors in requirements cannot be removed in software development. In this paper, we present a ripple effect analysis method in software requirements modification for data flow requirements. This method enables to specify areas of ripple effects of data flow diagrams in deletion and/or modification of requirements. Our method will be illustrated with examples using a prototype system based on the method.

Keywords: Data flow requirements
Ripple effect analysis in requirements modification · Data flow diagram

1 Introduction

Software requirements may change and evolve in software development. Such a requirement change and evolution may occur ripple effects and different requirements propagated by the original requirements change should be modified. Without proper impact analyses of requirements changes, requirements modification caused by ripple effects cannot be performed.

In this paper, we will introduce a ripple effect analysis method in requirements changes. We focus on data flow diagrams (DFDs), because DFDs are widely used in practical software development so far. DFD is a main diagram in Structured Analysis [3,10]. Data dictionary and process description are sub models in Structured Analysis. We use these three models in ripple effect analysis in data flow requirements modification. We regard deletions and/or updates of requirements as requirements changes. We do not apply ripple effect analysis method to data stores and actors in data flow diagrams, because we assume that data stores and actors are out of the target system.

2 Related Work

In [13] a ripple effect method of requirements evolution using relationship matrix among requirements is proposed. In [12] another ripple effect method of requirements evolution using relationship matrix among requirements is proposed. In

M. Virvou et al. (Eds.): JCKBSE 2018, SIST 108, pp. 1–11, 2019.
https://doi.org/10.1007/978-3-319-97679-2_1

case of data flow requirements, reachable dataflow requirements are not always affected by requirement change. So, their method cannot be applied to ripple effect analysis of data flow diagram.

Impact analysis methods of requirements changes are widely researched and proposed [1,2,4–6,8,9,11]. However, these methods cannot be applied to data flow requirements. Our method can be applied to data flow requirements and detect requirements to be changed through ripple effect analysis.

3 Ripple Effect Analysis Method

In structured analysis, data flow diagram, data dictionary and process description are modeled as a result of analysis. In this section we introduce rules of data flow diagram and states of data flow/process in ripple effect analysis.

3.1 Rules of Data Flow Diagram

We assume that data flow diagram should keep the following rules

- Each process should have both at least one input and at least one output.
- Data store and actor have at least one data flow.
- Each data flow should be connected with at least one process.

3.2 States of Data Flow in Ripple Effect Analysis

We assume four states of data flow in ripple effect analysis. These are I (initial state), N (No effect), C (Change), and D (Delete).

- Data flow with state "I" means that ripple effect analysis has not been applied to this data flow yet.
- Data flow with state "N" means that this data flow has no effect.
- Data flow with state "C" means that this data flow should be changed.
- Data flow with state "D" means that this data flow should be deleted.

3.3 State of Process in Ripple Effect Analysis

We assume six states of process in ripple effect analysis. These are I (initial state), N (No effect), CC (Change/Change), CN (Change/No effect), NC (No effect/Change) and D (Delete).

- Process with state "I" means that ripple effect analysis has not been applied to this process.
- Process with state "N" means that this process has no effect.
- Process with state "CC" means that both input and output of this process should be changed.
- Process with state "CN" means that input of this process should be changed, but output has no effect.
- Process with state "NC" means that output of this process has no effect, but input should be changed.
- Process with state "D" means that this process should be deleted.

Table 1. State of process decided by states of input data.

No.	Input	Process
1	N	Check the lower DFD
2	D	D
3	Not D nor not N	Check the lower DFD

3.4 Assumptions of Ripple Effect Analysis

We assume the following conditions.

1. Changed requirements are deletion and/or update of process(es)/data flow(s) in DFD.
2. DFDs, DDs, and process descriptions before changing requirements are correctly specified, that is to say, they obey the rules described in Sect. 3.1.
3. Each process in the bottom level of DFDs has process description.
4. Data flows and processes change their states into deletion, update, or no effect by ripple effect analysis.

3.5 Procedure of Ripple Effect Analysis

Before analysis, an analyst identifies deleted/updated/no effected processes/data flows. Deletion/update of processes/data flows correspond to changed requirements. Analyst may specify no changed processes/data flows separated from changed requirements as no effected processes/data flows.

The procedure of ripple effect analysis will be shown below.

Step 1: When data flows are deleted/updated specified as changed requirements, processes produce/receive these data flows are identified.
Step 2: Using Tables 1, 2, 3, 4 and 5, states of elements of DFD will be decided.
Step 3: If not decided, lower DFD will be analyzed. If lower layer is the bottom layer, analyst decides states of processes/data flows by interpreting process descriptions. The states decided by analyst will be transferred to states of DFD elements of the upper layer.
Step 4: Step 2 and 3 will be repeated until there exist no more ripple effects.
Step 5: If there exist any inconsistent states of elements of DFD, analysis will be stopped.

Table 1 shows that state of process can be decided by the states of input data. The second column shows that if states of all of input data are deleted, the state of process that receives the input becomes deleted. Otherwise, the state of process cannot be decided and the lower DFD corresponding to the process should be checked. This table is used when the state of input data are decided or specified, but the state of process that receives the input data is not fixed. As for checking the lower DFD, suppose a process labeled "...X" that consists of three sub-processes labeled "...X.1," "...X.2," and "...X.3" respectively in the lower DFD as shown in Fig. 1.

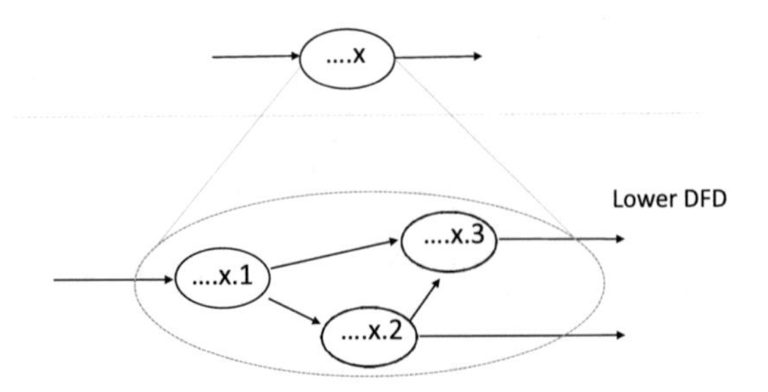

Fig. 1. A process and its lower DFD.

Table 2. State of process decided by states of output data.

No.	Output	Process
4	N	Check the lower DFD
5	D	D
6	Not D nor not N	Check the lower DFD

If the states of all the sub-processes are "Ds," the state of the parent process should be "D." If the states of all the sub-processes are "Ns," the state of the parent process should be "N." Otherwise, the state of the parent process should be "CC", "CN," or "NC."

In case of data structure defined in Data Dictionary, when the states of all the data elements are "Ds," the state of parent data should be "D." When the states of all the data elements are "Ns," the state of parent data should be "N." Otherwise, parent data should be "C."

Table 2 shows that the state of process can be decided by the states of output data. The second column shows that if states of all of output data are deleted, the state of process that generates the output becomes deleted. Otherwise, the state of process cannot be decided and the lower DFD corresponding to the process should be checked. This table is used when the state of output data are decided or specified, but the state of process that generates the output data is not fixed.

Table 3 shows that the state of output data can be decided by both the state of input and the state of process. The first column shows that if states of input data are deleted and if state of the process that receives the input is deleted, then state of output that generated by the process becomes deleted. The second column shows that if states of input data are deleted and if state of the process that receives the input is not deleted, then it becomes inconsistent between the input and the process, and analysis will be stopped.

Table 4 shows that state of input data can be decided by both the state of output and the state of process. The first column shows that if states of output data are deleted and if state of the process that generates the output is deleted, then state of input that received by the process becomes deleted. The second column shows that if states of output data are deleted and if state of the process that generates the output is not deleted, then it becomes inconsistent between the output and the process, and analysis will be stopped.

Table 3. States of output decided by states of input and process.

No.	Input	Process	Output
7	D	D	D
8	D	Not D (inconsistent)	–
9	N	N	N
10	N	NC	Check the lower DFD
11	N	D or CN or CC (inconsistent)	–
12	Not D nor not N	CN	N
13	Not D nor not N	CC	Check the lower DFD
14	Not D nor not N	D or NC or N (inconsistent)	–

Table 4. States of input decided by states of output and process.

No.	Output	Process	Input
15	D	D	D
16	D	Not D (inconsistent)	–
17	N	N	N
18	N	CN	Check the lower DFD
19	N	D or NC or CC (inconsistent)	–
20	Not D nor not N	NC	N
21	Not D nor not N	CC	Check the lower DFD
22	Not D nor not N	D or CN or N (inconsistent)	–

Table 5. State of process decided by states of input and output.

No.	Input	Output	Process
23	N	N	N
24	N	Not D nor not N	NC
25	Not D	D (inconsistent)	–
26	D	D	D
27	D	Not D (inconsistent)	–
28	Not D nor not N	N	CN
29	Not D nor not N	Not D nor not N	CC

Table 5 shows that state of process can be decided by both the state of input and the state of output. The first column shows that if states of input data and output data are not effected, then state of process that receives the input and generates the output becomes not effected. The second column shows that if states of input data are not effected and if the states of output data are not N nor not D, then state of process that receives the input and generates the output becomes NC. The third column shows that if the states of input data are not D and if the states of output data are D, then it becomes inconsistent between the input and the output, and analysis will be stopped.

Table 6 is used, when the state of process is specified, but states of both input and output of the process are not fixed.

Table 6. States of input and output decided by state of their process.

No.	Process	Input	Output
30	N	N	N
31	D	D	D
32	CN	Check the lower DFD	N
33	NC	N	Check the lower DFD
34	CC	Check the lower DFD	Check the lower DFD

4 Example: Business Trip Expenses

Suppose a system of business trip expenses. Originally transportation expenses, accommodation fee, and daily allowance are paid as trip expenses, and we give a changed requirement that daily allowance will not be paid. In Fig. 2, a business tripper sends his trip information to a receptionist, and the receptionist sends the information to accountant, and then final trip expenses will be paid to the business tripper. In this figure, business tripper is modeled as an actor and receptionist and accountant are modeled as processes.

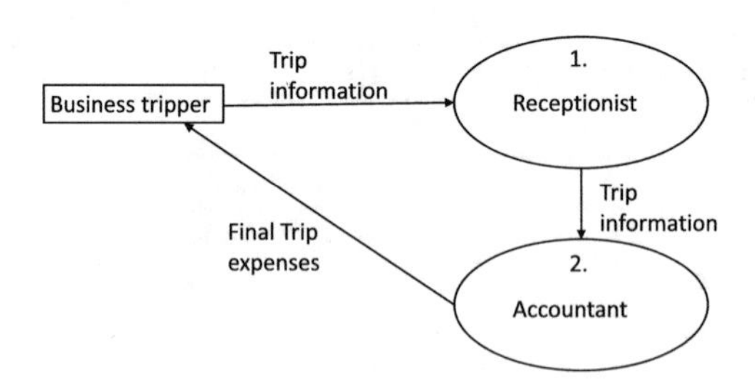

Fig. 2. DFD of system of business trip expenses.

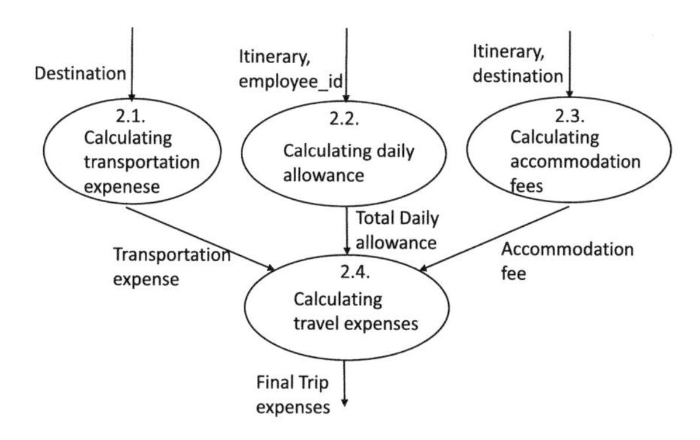

Fig. 3. Detailed DFD of process labeled "Accountant."

Figure 3 is a lower DFD of the process labeled Accountant. In this figure, transportation expense, accommodation fee, and daily allowance are paid as trip expenses. The "trip information" in Fig. 2 consists of "destination," "itinerary," and "employee_id." We delete "total daily allowance" in this system as a changed requirement as shown in Fig. 4.

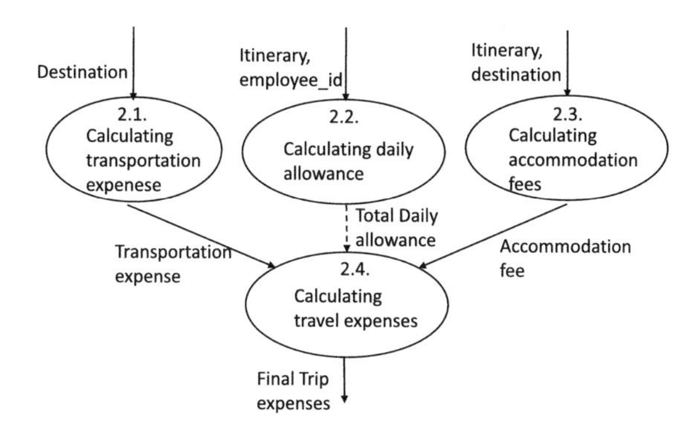

Fig. 4. DF labeled "Total daily allowance" is deleted.

According to the 2nd column of Table 2, the process labeled "2.2 Calculating daily allowance" becomes deleted. According to the first column of Table 4, two input data flows of the process become deleted. That is to say, "Itinerary and employee_id" become deleted.

As for the process "2.4," according to the 3rd column of Table 1, the lower DFD of this process or process description should be checked. In Fig. 5, process description of this process is shown. By investigating this description, analyst decided the state of this process. In this case, input data of the process are changed, because daily allowance is deleted, but final trip expenses need not be changed. So, analyst decided the state of this process as "CN." Processes labeled 2.1 and 2.3 have no effects. In Fig. 6, italic characters show states of DFD elements.

Trip expenses is a sum of transportation expense, daily allowance, and accommodation fee.

Fig. 5. Process description of "Calculating travel expenses."

These results will be used for interpreting states of the DFD elements of upper DFD shown in Fig. 2. In this example, "trip information" in Fig. 2 consists of "Destination," "Itinerary," and "employee_id" in Fig. 3. According to Fig. 6, "Destination" and "Itinerary" have no effects, but "employee_id" is deleted, so "trip information" should be "C" (Changed). Since sub-processes 2.1, 2.3 have no effects, but sub-processes 2.2 and 2.4 are deleted, so the upper process labeled 2 should be changed. The process labeled "1" in Fig. 2 should be checked the lower DFDs, because of the 3rd column in Table 2. In this case, analyst decided that the process "1" had no effects by checking the lower DFDs (not specified in this paper for the space limitations.) Finally, states of DFD elements of Fig. 2 are shown in Fig. 7.

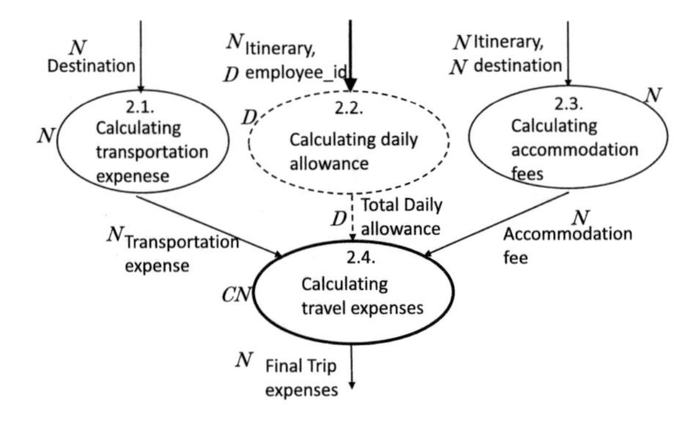

Fig. 6. States of detailed DFD elements of process labeled "Accountant."

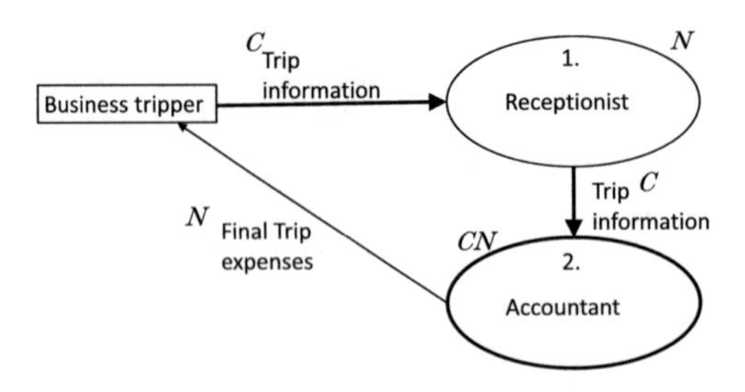

Fig. 7. States of DFD elements of Fig. 2

We have developed a prototype system based on the proposed method. This system is written with Java standard edition 1.8 using Eclipse Neon.3 release (4.6.3). The line of source code is about 3,600. This system is a 6 person-month product.

We adopted iEdit [7] as a DFD editor/viewer. First users should make DFDs with iEdit. Second, our system reads internal representation of DFDs produced by iEdit and provides the users with names of DFD elements so that the users can select no effect/changed/deleted DFD elements as changed requirements.

After specifying changed requirements, our system starts to analyze ripple effects using tables in Sect. 3.5. When reaching DFDs of the bottom layer, system shows process description of each process of DFDs of the bottom in order that the users can decide states of DFD elements of the bottom layer.

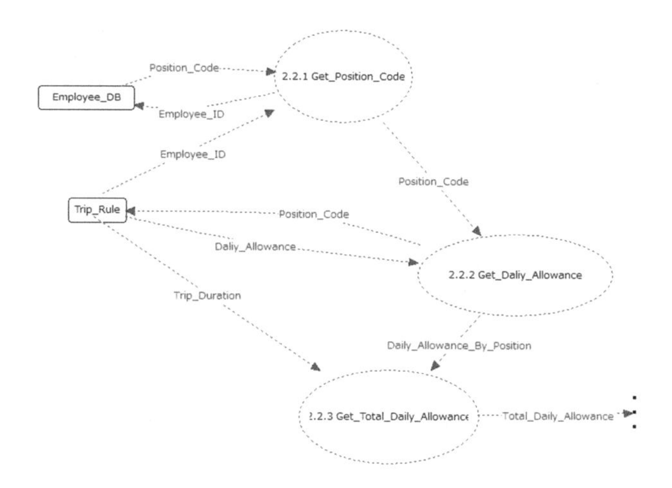

Fig. 8. The bottom layer of DFD of calculating daily allowance.

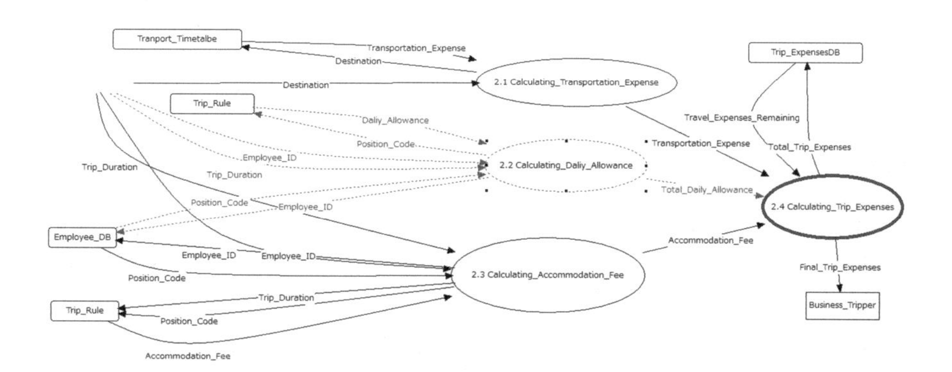

Fig. 9. The third layer of DFD of accountant process.

Lastly states of DFD elements of layers will be provided as shown in Figs. 8, 9, and 10. In these figures, processes and data flows of dotted lines mean DFD elements of "D" states. Process of thick line and data flow of thick line mean DFD elements of "C (CC, CN, NC)" states.

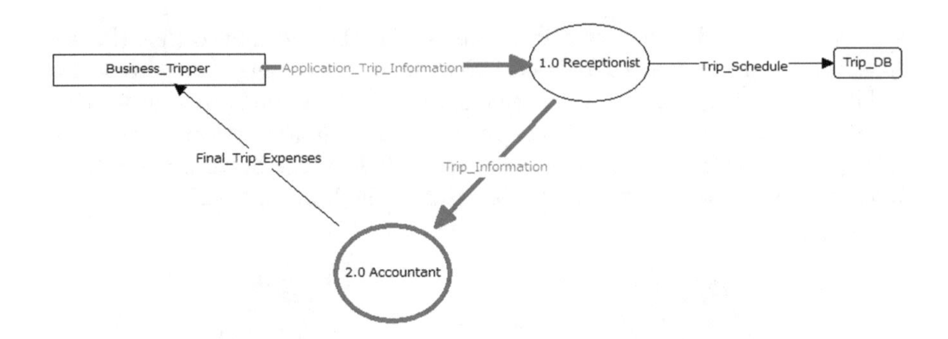

Fig. 10. The second layer of DFD of trip expenses system.

Comparing ripple effect analysis by an experienced software analyst with analysis by the prototype system, both the precision and recall ratios are 100%. So, our method can perfectly make ripple effect analysis in this example.

5 Conclusion

We have developed a ripple effect analysis method for data flow requirements. Our method enables to detect DFD elements in DFDs to be deleted/changed. We have developed a prototype system based on the method. Through an experiment, we confirmed the effectiveness of our method and system.

We regard deletion and update of DFD elements as requirements changes. We have to enhance our method by introducing addition of requirements as requirements changes. Evaluations of our method/system by applying them to practical DFDs and data flow requirements changes are another future work.

References

1. Arora, C., Sabetzadeh, M., Goknil, A., Briand, L.C., Zimmer, F.: Change impact analysis for natural language requirements: an NLP approach. In: Proceedings of the 2015 IEEE 23rd International Requirements Engineering Conference (RE15), pp. 6–15 (2015)
2. Briand, L.C., Labiche, Y., O'Sullivan, L.: Impact analysis and change management of UML models. In: Proceedings of the International Conference on Software Maintenance (ICSM 2003), pp. 1063–6773 (2003)
3. DeMarco, T.: Structured Analysis and System Specification. Prentice-Hall Inc., San Francisco (1979)

4. Gezgin, T., Henkler, S., Stierand, I., Rettberg, A.: Impact analysis for timing requirements on real-time systems. In: Proceedings of the 2014 IEEE 20th International Conference on Embedded and Real-Time Computing Systems and Applications, pp. 1–10 (2014)
5. Haleem, M., Rizwan Beg, M.: Impact analysis of requirement metrics in software development environment. In: Proceedings of the 2015 IEEE International Conference on Electrical, Computer and Communication Technologies (ICECCT), pp. 1–6 (2015)
6. Hassine, J., Rilling, J., Hewitt, J., Dssouli, R.: Change impact analysis for requirement evolution using use case maps. In: Proceedings of the 2005 Eighth International Workshop on Principles of Software Evolution (IWPSE 2005), pp. 81–90 (2005)
7. iEdit. http://kondoumh.com/software/iedit.html
8. Mokammel, F., Coatanea, E., Bakhouya, M., Christophe, F., Nonsiri, S.: Impact analysis of graph-based requirements models using PageRank algorithm. In: Proceedings of the IEEE International Systems Conference (SysCon), pp. 731–736 (2013)
9. Saito, S., Hoshino, T., Takeuchi, M., Hiraoka, M., Kitani, T., Aoyama, M.: Empirical analysis of the impact of requirements traceability quality to the productivity of enterprise applications development. In: Proceedings of the 2012 19th Asia-Pacific Software Engineering Conference (APSEC 2012), pp. 330–333 (2012)
10. Stevens, W., Myers, G., Constantine, L.: Structured design. IBM Syst. J. **13**(2), 115–139 (1974)
11. Suma, V., Shubhamangala, B.R., Manjunatha Rao, L.: Impact analysis of volatility and security on requirement defects during software development process. In: Proceedings of the International Conference on Software Engineering and Mobile Application Modelling and Development (ICSEMA 2012), pp. 1–5 (2012)
12. Yu-Qing, Y., Shi-Xian, L. Xian-Ming, L.: Quantitative analysis for requirements evolution's ripple-effect. In: Proceedings of the 2009 International Asia Conference on Informatics in Control, Automation and Robotics, pp. 423–427 (2009)
13. Zhao, J., Liu, J., Feng, W.: Study on the ripple effects of requirement evolution based on feedforward neural network. In: Proceedings of the International Conference on Measuring Technology and Mechatronics Automation 2010, pp. 677–680 (2010)

Analysis of Specification in Japanese Using Natural Language Processing

Kozo Okano[1(✉)], Kazuma Takahashi[1], Shinpei Ogata[1], and Toshifusa Sekizawa[2]

[1] Faculty of Engineering, Shinshu University, 4-17-1 Wakasato, Nagano 380-8553, Japan
okano@cs.shinshu-u.ac.jp

[2] College of Engineering, Nihon University, 1 Nakagawara Tokusada Tamuramachi, Koriyama 963-8642, Japan

Abstract. A requirement specification for software is usually described in a natural language and thus may include sentences containing ambiguity and contradiction. Problems due to the ambiguity often occur at the stage of the verification process of software development, and this forces developers to go back to the design process again. In order to prevent this kind of rework, a method of automatically converting a required specification written in Japanese to a state transition model is desired to help detect ambiguity and contradiction points of the specification. This paper proposes a method for this purpose, and reports on the result of applying the method to a specification example of an electric pot.

1 Introduction

Most software specifications are written in a natural language, which can result in ambiguous descriptions of what these specifications really mean [1]. Such ambiguity can sometimes make rework of development of software necessary. Ambiguity and incorrectness of specifications are sometimes found by formal approaches including model checking techniques. However, in order to perform such checks, we have to prepare in advance a formal specification of a target system [2]. In an earlier work, we proposed a method that semi-automatically generates a state transition model of a target system from a given specification of the system written in Japanese [3] which succeeds in extracting state variables from a given specification, but it does not propose how to extract transitions from the specification. This paper newly proposes a method to extract transitions from a given specification. This paper describes our approach and experimental results obtained by applying our method to a specification of an electric pot [4] that is described by SESSAMI. The experimental results show that our method has good recall rate performance.

This paper is organized as follows. Section 2 introduces some past techniques and related work as background. Section 3 outlines our proposed approach. Sections 4 and 5 present experimental evaluation and discussion, respectively. Finally, Sect. 6 concludes the paper.

M. Virvou et al. (Eds.): JCKBSE 2018, SIST 108, pp. 12–21, 2019.
https://doi.org/10.1007/978-3-319-97679-2_2

"*Mori*" 'forest'	***noun***	***noun phrase***	***verb phrase***	***noun phrase***	***noun phrase***	***statement***
	"*de*" 'at'	***particle***				
		"*Minotta*" 'ripe'	***verb***	***noun phrase***	***noun phrase***	***verb phrase***
			"*ringo*" 'apple'	***noun***	***noun phrase***	***verb phrase***
				"*wo*" 'to'	***particle***	
					"*tabeta*" 'ate'	***verb***

Fig. 1. Parsing a sentence based on a given grammar using the CYK Algorithm.

2 Background

2.1 Related Techniques

The CYK Algorithm [5] composes grammar trees from a given context-free grammar G and a given sentence s using a dynamic programming algorithm. Usually the grammar G is given in Chomsky-normal form. It requires $O(\mid s\mid^3 \cdot \mid G \mid)$ calculation time. Figure 1 shows an analysis example of CYK algorithm. For a sentence "*Mori de minotta ringo wo tabeta*"[1] morphological analysis is performed and each morpheme is identified. For these morphemes and the given typical Japanese grammar, CYK algorithm is performed.

For example, in Fig. 1, the noun "*mori*" is adjacent to the particle "*de*" thus "*mori de*" becomes a noun phrase. Such analysis is performed for all morphemes. When a sentence is produced the analysis succeeds.

Kuromoji [6] is an open source Japanese morphological analysis tool written in Java. Morphological analysis determines the parts of speech of the components of a given text, in other words, the morphemes, which are the minimum units of a text, using its grammar and a dictionary with information on parts of speech. Mecab [8] and JUMAN [9] are famous tools for Japanese morphological analysis. We, however, use Kuromoji because of its ease of use with Java.

2.2 Related Work

Natural Language Processing (NLP) is an important technique in information processing. The recent development of NLP has been remarkable [7] and there are many applications of it to various fields. For Japanese texts, KNP [10] and Cabocha [11] are famous tools for syntax analysis. Some of them use support vector machines (SVMs) for processing.

Huy and Onishi [12] have proposed a method for checking the correctness of requirements based on ontology and NLP. The group has also studied a method to generate scenarios from a domain specific language (DSL) based on a restricted natural language [13]. Yoshiura and Nakanishi, *et al.* [14] have proposed a method based on a Japanese unification grammar to extract simple sentences from technical documents to generate sentences that assist in describing failure modes for HAZOP/FMEA [15].

[1] The actual sentence is given in pure Japanese which uses Chinese characters, Hiragana, and Katakata without space-delimiters. The same applies hereafter.

3 Proposed Method

Our proposed method consists of the following six steps. Each step is performed on a sententece, therefore our proposed method does not deal with cotext of sentences.

1. Perform morphological analysis on required specifications written in Japanese using existing tools;
2. Perform parsing using the CYK method;
3. Output the syntax analysis result in XML;
4. Classify outputs;
5. Extract state variables; and
6. Extract state transitions.

3.1 Morphological Analysis

From a target requirements document, we manually extract each request specification and rewrite it to a sentence so that the syntax can be classified. After that, we reformulate the sentences using morphological analysis using Kuromoji.

3.2 Parsing

For each sentence, we perform parsing using the results of morphological analysis and tagging of parts of speech which is one of Kuromoji's functions.

3.3 XML Output of Morpheme Analysis and Parsing Results

After the morphological analysis and parsing of a sentence succeeds, the result is output in XML format. Listing 1.1, shows the result of an example sentence "*Suiisensa ha, suii wo kenchisuru.*" The morphological analysis result is in each of `<tokens/>` tags and the parsing result is in the `<nounphrase/>` and `<verbphrase/>` tags.

Listing 1.1. Analysis Result in XML

```
<tokens>
  <word>Suii</word>
  <pos>noun</pos>
</tokens>
<tokens>
  <word>sensa</word>
  <pos>noun</pos>
</tokens>
<tokens>
  <word>ha</word>
  <lemma>ha</lemma>
  <pos>particle</pos>
</tokens>
<tokens>
  <word>, </word>
  <pos>symbol</pos>
</tokens>
```

```
<ommitted/>

  <nounphrase>
    <word pos="1">suiisensaha</word>
    <part>noun phrase</part>
    <word pos="2">,</word>
    <part>symbol</part>
  </nounphrase>
  <verbphrase>
    <word pos="01">suiiwo</word>
    <part>noun phrase</part>
    <word pos="02">kenshutusuru</word>
    <part>verb phrase</part>
  </verbphrase>
  <verbphrase>
    <word pos="01">sensaha,</word>
    <part>noun phrase</part>
    <word pos="02">suiiwokenshutusuru</word>
    <part>verb phrase</part>
  </verbphrase>
note: the element of <word/> is actually Japanese words
```

3.4 Classification of Sentences

From the combination of clauses and phrases, we classify the sentences into three types: "definition," "processing," and others. Using the information on the classification, we retrieve nouns as state variable candidates [16].

Classification Method. The classification of sentences is performed in the following two steps.

1. Split sentences into condition clauses and action clauses from the parsing result.
2. Judge by connections and pattern of the parts of speech of the condition clause and the action clause.

For a sentence "*Botan ga osareta toki, taima wo teishi suru,*" it is divided into "*Botan ga osareta toki,*" and "*taima wo teishi suru,*" from the result of the parsing. In this case, the first clause is the condition clause, and the second is the action clause.

Processing Sentences. Focusing on the condition clause, when the following conditions apply, a sentence is judged to be a processing sentence.

- If a verb or auxiliary verb + noun ("*toki*", "*baai*") phrase is included in the sentence.
  (ex.) "*Botan ga osareta toki, taima wo teishi suru.*"
- If a verb or auxiliary verb + particle ("*to*") is included in the sentence
  (ex.) "*Futa ga akerareru to, dengen ga off ni naru.*"

Definition Sentence. Sentences which cannot be matched by the Processing classification rule and only consist of noun + particle (ha) are categorized as definition sentences.

(ex.) "*Suiisensa ha, suii wo kenchisuru.*"

3.5 Extraction of State Variables

We extract state variables using the results of the classification. The extraction of state variables is performed in different ways for definition and processing sentences.

In the case of a definition sentence, the noun first used as a sentence's subject is extracted as a candidate for a state variable. If a noun is also behind a noun that is being used for the first time, it is also taken as a noun. For the sentence "*Suiisensa ha, suii wo kenchisuru,*" the noun used first is "*suii*" and the word after it is also a noun "*sensa*". In this case, the candidate for the state variable of this sentence is "*suii-sensa*".

In the case of the processing sentences, the first noun used in the subject of the condition clause and the first noun in the action clause are the candidates. In the case of processing, if there is a noun after a noun, it is also taken as a noun. For a sentence "*Taima botan ga osareta toki, taima wo kidou suru,*" both "*taima botan*" used at the beginning of the condition clause and "*taima*" used in the action clause are candidates for state variables.

3.6 Extraction of State Transitions

Every state transition is extracted from a processing sentence. Sentences in natural language do not directly represent events or actions [17]. Therefore, it is necessary to modify sentences of natural language in order to extract state transitions. Here we assume that each processing sentence is in the form of a sentence which explicitly includes conditions, the event, the previous state, and the post state. A typical sentence is shown below.

"*(state)* no toki, *(condition)* baai *(action)* wo okonau to *(state)* ni naru." In English it would be "When *(state)* and *(condition)* holds, *(action)* to *(state)*."

4 Evaluation

4.1 Creating Evaluation Criteria (Correct Answer Set)

Based on a document [4], four novices of UML consulted and created a state transition diagram shown in Fig. 2. The main part is derived using only the description of Chap. 7 of the document [4]. They also referred to the descriptions of some other chapters. In this way, we obtained the state transition diagram. We use this diagram as the correct answer set for the experiment.

The natural language description used totaled 33 sentences. The state variables are identified from Fig. 2.

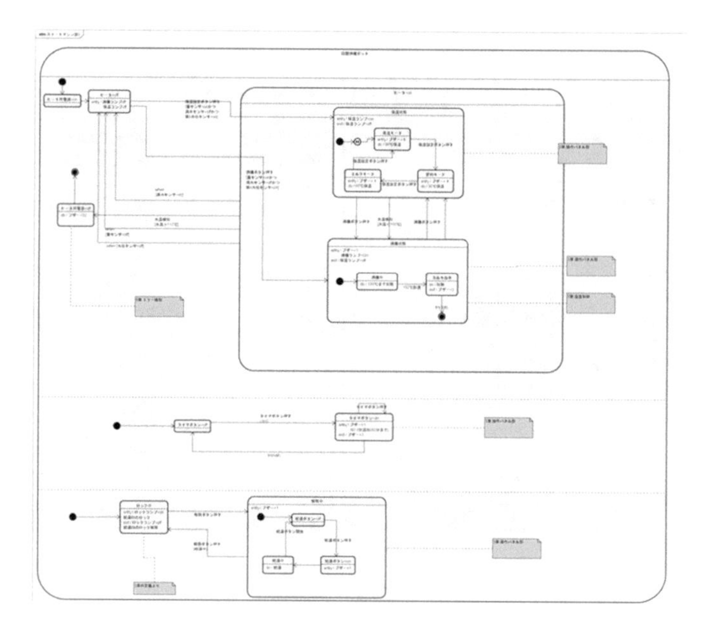

Fig. 2. State diagram derived manually.

4.2 Preformatting the Input Sentences

Because there are compound sentences and the condition clause is implicitly suggested, we formatted each of them as a set of single sentences.

4.3 State Variables

Table 1 shows the results of extracted state variables and the correct answers. The number of state variables in the correct answer set created from Fig. 2 is 26, whereas the number of state variables extracted by the proposed method is 35. There are 26 matches with the state variables in the correct answer set. The recall and precision of the proposed method are 1.00 and 0.74, respectively. Note that we regard the word 'lid' "*futa*" as the same word as 'lid sensor' "*futa sensa.*"

4.4 Extract State Transitions

Tables 2 and 3 show the results of the correct answer set and extracted state transitions, respectively. The number of state transitions extracted by the proposed method is 33, while 19 state transitions are the correct answers. We call a transition pair with every element matched a *perfect match*, where an element is either a previous state, a condition, an action, or a post state. Also, we call a transition pair with some element matched a *partial match*. The number of perfect matches is 9, while the number of partial matches is 18. Table 4 shows the recall and precision of the proposed method.

Table 1. State variable names: correct set and extracted set

Correct Answer	Extracted Variable Name
n Water Level Sensor	n Water Level Sensor
Full Water Level Sensor	Full Water Level Sensor
Lid Sensor	Lid
Heater	Heater
Power Supply for Heater	Power supply for Heater
Water Temperature	Water Temperature
Heat Retaining setting Button	Heat Retaining setting Button
Timer Button	Timer Button
Boil Button	Boil Button
Unlock Button	Unlock Button
Hot Water Supply Button	Hot Water Supply Button
High Temperature Mode	High Temperature Mode
Milk Mode	Milk Mode
Save Mode	Save Mode
Timer	Timer
Buzzer	Buzzer
Heat Retaining Lamp	Heat Retaining Lamp

Correct Answer	Extracted Variable Name
Boiling Lamp	Boiling Lamp
Lock Lamp	Lock Lamp
Heat Retaining Mode	Heat Retaining Mode
Boil	Boil
Hot Water Supply	Hot Water Supply
lock	lock
Chalk Removal	Chalk Removal
Hot Water Inlet	Hot Water Inlet
	Boiling State
	Pump
	Water
	Boil
	Heat Retaining State
	Heating
	Pot
	hand
	heating

Note: we translate Japanese words to English ones.

Table 2. Correct transitions

Previous state	Condition	Action	Post state
Heater off	Lid Sensor on, Full Water Level Sensor off	press HPSB	Heat Retaining Mode
Heater off	Lid Sensor on, Full Water Level Sensor off	press BB	Boiling Mode
High Temperature Mode		press HPSB	Save Mode
Save Mode		press HPSB	Milk Mode
Milk Mode		press HPSB	High Temperature Mode
Heat Retaining Mode		press BB	Boiling Mode
Boiling Mode		press BB	Heat Retaining Mode
Heat Retaining Mode	Water Temperature $< 100^{\circ}$	sense water temperature	Boiling Mode
Heater on	Water Temperature $> 110^{\circ}$	sense water temperature	Power Supply for Heater off
Timer Button off		press TB	Timer Button on
Timer Button on	timeout		Timer Button off
Timer Button on		press TB	Timer Button on
Hot Water Inlet lock		press CB	Hot Water Inlet unlock
Hot Water Inlet lock	not water supplying	press UB	Hot Water Inlet lock
Hot Water Supply off		press WSB	Hot Water Supplying
Hot Water Supplying		release WSB	Hot Water Supply off
Heater on	Lid Sensor off		Heater off
Heater on	n Water Level Sensor off		Heater off
Heater on	Full Water Level Sensor on		Heater off

Note: we translate Japanese words to English ones.
HPSB: Heat Preservation Setting Button, TB: Timer Button, BB: Boil Button, WSB: Hot Water Supply Button, UB: Unlock Button

Table 3. Extracted transitions

Previous state	Condition	Action	Post state
High Temperature Mode		press HPSB	Save Mode
Save Mode		press HPSB	Milk Mode
Milk Mode		press HPSB	High Temperature Mode
Hot Water Inlet lock		press UB	Hot Water Inlet unlock
WSB pressing		release UB	Hot Water Supply stop
Boiling		press BB	Heat Heat Retaining State
Boiling		press BB	Heat Retaining
Heat Retaining		press BB	Boiling
Hot Water Inlet unlock		press UB	Hot Water Inlet lock
Boiling State	Water Temperature reaches 100°	Heat 3 min.	Heat Retaining
		press TB	Timer start
		press TB	increase Timer 1 min.
		press WSB	Hot Water emitting
		finish Boiling	Boiling Lamp off
		open Lid	Heater stop
		open Lid	BB stop
		press TB	Buzzer on
		press HPSB	Buzzer on
		press BB	Buzzer on
		press UB	Buzzer on
		press WSB	Buzzer on
		timer is timeout	Buzzer on three times
		finish Boiling	Buzzer on three times
		press BB	do Chalk Removal
pressing WSB			Hot Water Supplying
Boiling			Boiling
Water Temperature ≥ 110°			Power Supply for Heater off
Water Temperature ≥ 110°			Buzzer on 30 sec.
n Water Level Sensor on, Full Water Level Sensor off			control enable
Stop			Boiling Lamp off
Stop			Heat Retaining Lamp off
Hot Water Supplying			lock disable
Heat Retaining Mode	not 100°		Boiling

Note: we translate Japanese words to English ones.
HPSB: Heat Preservation Setting Button, TB: Timer Button, BB: Boil Button, WSB: Hot Water Supply Button, UB: Unlock Button

5 Discussion

5.1 State Variable

Our method can extract state variables without omission. The number of extracted states by our method is about 1.3 times the number of correct answers. The reasons for this are as follows.

- Extraction of nouns with similar meaning.
- Extraction of nouns whose state cannot be determined in the system.

Nouns "*kanetsu*" (heating) and "*futtou*" (boiling) are examples of the first. Nouns "*mizu*" (water) and "*te*" (hand) are examples of the latter. If these nouns can be excluded, higher precision can be expected.

Table 4. Recall rate and precision rate

	Recall	Precision
State variable	1.00	0.74
Transition perfect match	0.47	0.27
Transition partial match	0.95	0.55

5.2 Transition

The recall rate of partial matching demonstrates good performance as high as 0.94. The reason why the recalls of perfect matches is low is that states and events are sometimes not described explicitly in the specification. Therefore, if we can identify the contexts, the performance will be improved.

In addition, from the results we find that, for a transition T which is extracted from a sentence s,

- if an event is missing in s, the word estimated as the previous state in T can often be the condition of T; and
- if the previous state is missing in s, T can often be an "entry action."

For example, for "*Suion ga 100° ijo no tokini, 30 byo buza wo narasu,*" our method decides that "*Suion ga 100° ijo*" is a previous state. However, this phrase is a condition. Also, for "*Taima botan ouka wo okonau to buza wo narasu,*" our method decides that "*buza wo narasu*" is an event of the transition. The transition, however, should be classified as an entry action. Based on these findings, we can improve the precision.

6 Conclusion

Ambiguity and incorrectness of specifications are sometimes found by formal approaches including model checking techniques. In order to perform such formal approaches, we have to prepare in advance a formal specification of a target system. We proposed a method that semi-automatically generates a state transition model of a target system from a given specification of the system written in Japanese. The experimental results showed that our method has good performance at the recall. We will try to extract state transition diagram automatically. To deal with cotext of sentences is also included in future work.

Acknowledgements. Part of this research is supported by Grants-in-Aid C16K00094. The authors would like to thank Mr. Yusuke Naka and Mr. Yusuke Sugiyama for deriving the UML state machine diagram which served as the base of the correct answer set. Authors also thank Prof. Takahiro Yamada of JAXA for valuable comments.

References

1. Meyer, B.: On formalism in specifications. IEEE Softw. **2**(1), 6–26 (1985)
2. Tripathy, A., Rath, S.K.: Object oriented analysis using natural language processing concepts: a review. arXiv:1510.07439vl [cs.SE]26 (2015)
3. Okano, K., Takahashi, K., Naka, Y., Ogata, S., Sekizawa, T.: Analysis of specification in Japanese using natural language processing and review supporting with speech synthesis, IEICE Technical report, vol. 117, no. 465, KBSE2017-52, pp. 79–84 (2018). (in Japanese)
4. SESSAMI: embedded system educational material “Wadai-Futto pot GOMA-1015 type” required specification. http://www.sessame.jp/workinggroup2/POT_Specification.html. Accessed 11 Jan 2018. (in Japanese)
5. Kasami, T.: An efficient recognition and syntax-analysis algorithm for context-free languages, Scientific report AFCRL-65-758. Air Force Cambridge Research Lab, Bedford, MA (1965)
6. Kuromoji-Atilika-AppliedSearchInnovation. https://www.atilika.com/ja/products/kuromoji.html. Accessed 11 Jan 2018
7. Ranjan, N., Mundada, K., Phule, S., Phule, S.: A survey on techniques in NLP. Int. J. Comput. Appl. **134**(8), 6–9 (2016)
8. MeCab: yet another part-of-speech and morphological analyzer. http://taku910.github.io/mecab/. Accessed 11 Jan 2018
9. JUMAN. http://nlp.ist.i.kyoto-u.ac.jp/index.php?JUMAN. Accessed 11 Jan 2018
10. Sasano, R., Kurohashi, S.: A discriminative approach to Japanese zero anaphora resolution with large-scale lexicalized case frames. In: Proceedings of the 5th International Joint Conference on Natural Language Processing (IJCNLP 2011), pp. 758–766 (2011)
11. Kudoh, T., Matsumoto, Y.: Japanese dependency analysis based on support vector machines. In: Joint SIGDAT Conference on Empirical Methods in Natural Language Processing and Very Large Corpora EMNLP/VLC 2000 (2000)
12. Huy, B.Q., Ohnishi, A.: A verification method of the correctness of requirements ontology. In: Proceedings of the 10th Joint Conference on Knowledge-Based Software Engineering (JCKBSE 2012), pp. 1–10 (2012)
13. Ohnishi, A., Kitamoto, K.: A generation method of alternative scenarios with a normal scenario. IEICE Trans. Inf. Syst. **E93–D**(4), 693–701 (2010)
14. Yoshimura, K., Nakanishi, T., Ototake, H., Tanabe, T., Furusho, H.: Extraction of simple sentences from technical documents using Japanese unification grammar. IEICE Technical report, vol. 117, no. 465, KBSE2017-58, pp. 109–114 (2018). (in Japanese)
15. IEC 61882:2001: Hazard and operability studies (HAZOP studies) - Application guide
16. Yumikura, Y., Wada, T., Sumi, T., Fujinmoto, H., Murata, Y.: Applying natural language analysis to the evaluation of requirements specifications. IPSJ Technical report, 2013-SE-181, pp. 1–8 (2013). (in Japanese)
17. Tanaka, K., Ikeda, T., Deguchi, Y.: A software development system based on natural language specifications. In: Proceedings of the 48th National Convention of IPSJ, pp. 91–92 (1994). (in Japanese)

Deriving Successful Factors for Practical AI System Development Projects Using Assurance Case

Hironori Takeuchi[1(✉)], Shiki Akihara[2], and Shuichiro Yamamoto[3]

[1] Musashi University, Tokyo, Japan
h.takeuchi@cc.musashi.ac.jp
[2] IBM Japan, Ltd., Tokyo, Japan
shikiaki@jp.ibm.com
[3] Nagoya University, Nagoya, Japan
syamamoto@acm.org

Abstract. In this research, we considered projects to develop systems that use AI technologies including machine learning techniques for office environment. In many AI system development projects, both developers and users need to be involved in order to reach a consensus on discussion items before starting a project. To facilitate this, we propose a method of assessing an AI system development project by using an assurance case based on quality sub-characteristics of functionality to derive project success factors.

1 Introduction

With the recent rapid growth of machine learning technologies that include deep learning, now it is now possible to apply artificial-intelligence (AI) technologies to practical business uses. Many machine-learning-based AI programming modules, such as text classification and image recognition, have been developed and made available as application programming interfaces (APIs). As a result, developers do not have to worry about the details of the machine-learning algorithms but use the module function by just preparing the training data required for the machine-learning programming module. In this research, we consider system development projects using AI technology APIs.

Offices have started to apply AI technologies to support operators of inquiry services, answering queries about business operations, products or services, or screening operations using documents containing many types of individual data. When applying AI technologies to these business activities, we need to collect training data on the target business domain. In such situations, we need knowledge or experience on the domain. System developers from an IT vendor themselves cannot complete this activity, so members of client companies using developed systems need to be involved. Also, because the output of systems using machine learning technologies is probabilistic, we sometimes get a different output even if the input is slightly changed, and we get unexpected outputs for some

M. Virvou et al. (Eds.): JCKBSE 2018, SIST 108, pp. 22–32, 2019.
https://doi.org/10.1007/978-3-319-97679-2_3

inputs. Therefore, in a project, we need to define a performance threshold representing a degree of output correctness required in the target business. While developing a system, we iteratively update training data to improve performance or consider implementing a workaround plans for cases in which we cannot get expected results. Members from client companies also need to participate in discussion for these activities.

As described above, in projects developing a system using AI technologies, the client members involved need to understand their work items properly and provide their outcome artifact on time. Usually, before starting projects, each project member's work items and the project schedule are described in a project proposal, that all stakeholders agree to. However, sometimes project members from client's-side do not understand the importance of their work items and do not have enough time to provide high-quality artifacts such as training data. In such situations, the systems developed will not achieved the performance threshold agreed in project proposals. Also, client members might not understand the performance of output correctness in a business context or might not properly consider workarounds for cases in which they can get expected outputs for some inputs. As a result, client members are not satisfied with the system developed as a MVP and they will decide not to continue the project to the point of deploying it as a production system.

In this research, we consider these issues in system development project applying AI technologies to office business. Our research goal is to develop a metric to assess the readiness of AI system development projects. For this goal, we propose an assurance case based method to assess the projects. By applying the proposed method to real projects, we show that we can identify success factors for AI system development projects and validate that we can use the proposed method as a metrics to assess project readiness.

The rest of this paper is as follows. In Sect. 2 we describe related work. In Sect. 3, we introduce the AI system and describe the research hypothesis in this study. In Sect. 4, we describe the proposed method for assessing AI system development projects. In Sect. 5, we apply the proposed method to real project cases and show the effectiveness of the method. After the discussion in Sect. 6, we summarize the key points and future work in Sect. 7.

2 Related Work

A successful project is one in which we can achieve the project goal within budget and on time. Some researches [1,2] discuss the factors surrounding the success or failure of software projects. In [1], the relationship between project success and project features such as members, schedule or planned work items is investigated for business system development projects across industries. In [2], organization cultures for project success are investigated.

Project methodologies are proposed for projects using Big Data analytics or machine learning technologies. In [3], it is said that developing prototype systems iteratively is very useful for requirement elicitation and its validation.

Data analysis, the system development process, and the roles of project members (including data scientists) are discussed in [4,5]. These researches focus on starting actual projects and include a few studies on how to start projects properly. Though there is a general discussion on pre-conditions required for system development projects [6], concrete pre-conditions needed to start projects using AI technologies in particular are not discussed.

Assurance cases support the consensus building process by representing arguments with goal trees [7]. In an assurance case claims are decomposed using Goal Structuring Notation (GSN) [8]. Some patterns for developing assurance cases are proposed for various applications and the patterns' effectiveness is discussed [9]. Assurance cases can be applied to represent a project from a specific viewpoint. They have been mainly used in a specific domain for assurance that systems are safe or secure [10–12]. Also, for general IT projects, they are used for assessing the sufficiency of a requirements analysis or project risks [13–15]. However, they have not been used for assessing project readiness. In this research, we consider using assurance cases to assess the readiness of a project using AI technologies.

3 AI System and Research Hypothesis

3.1 AI System

In this research, we consider a practical AI project in which we developed a system containing AI technologies for an office environment. In an office, employees conduct various intelligent activities. It is said that there are three types of human intelligence: analytical intelligence, creative intelligence and practical intelligence [16]. In our research, we consider developing a system with analytical intelligence for offices that supports human activities or is substitute for them.

Analytical intelligence selects the optimal option from predefined ones as output for given input data [16]. In an office, this intelligence is used in daily activities such as inquiry service for service queries or business assessment based on documents. We can use machine learning technologies when we realize this intelligence as a software system. To use machine learning technology for system development, we need to define options on the target business domain and collect example inputs assumed for each option. A machine learning model is generated (Training) from training containing such pairs of options and examples. This model is deployed into a runtime machine learning engine, and the engine gets input data and provides output data using the model (Prediction). This system, called AI System, is illustrated in Fig. 1. In this research, we consider projects in which an AI system is developed in order to realize an analytical intelligence and to support business activities in an office.

3.2 Research Hypothesis

Usually, in software system development projects, system users are asked to participate in interview sessions for requirement elicitation, reviewing requirement

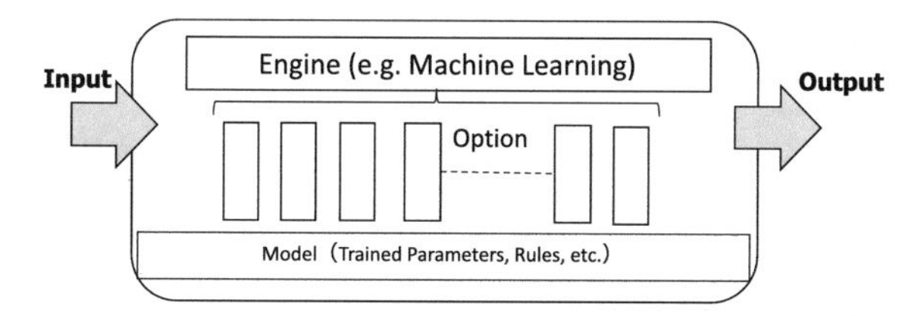

Fig. 1. AI system

specification documents and conducting user tests for the newly developed system. In addition to these activities, in AI system development projects, they are asked to develop training data. To apply a machine learning engine to a business domain and validate its sufficiency in business, the target domain's specific training data is mandatory. Therefore, training data is an important project artifact that users need to develop on schedule. Also, they are expected to discuss system performance, such as output suitability or usability for a real business from a business perspective.

Before starting a project, work items for both users and developers are defined in detail. The user side activities described above are listed and a schedule for the activities is agreed upon by users and developers. However, in AI system development projects, users do not understand the objectives of their activities or the significance of their artifacts because users and developers have a different understanding of the developed AI system. Therefore, sometimes users do not carefully consider the system's performance in a business context before staring the project or they cannot provide their artifacts (training data) by the appointed date. The result is a developed AI system that does not achieve the expected performance necessary business-wise and users are not satisfied with the system.

To solve this issue, before starting a project we propose a project assessment method based on an assurance case for AI system development projects. By developing assurance cases for AI system development projects, we can expect both users and developers to have a common understanding of the objectives and the significance of each work item assigned to them. We also expected to be able to assess the degree of attainment of the top goal by checking the evidence of sub-goals.

We define the following research hypotheses to show that we can solve practical issues in AI system development projects using our new method proposed in Sect. 4.

- We can identify successful factors on AI system development projects.
- We can use the method to form a project metric by which we can assess the maturity of the project preparation.

In an experiment in Sect. 5, we apply the proposed method to real projects and validate these hypotheses.

4 Proposed Method

In this section, we propose a method for assessing an AI system development project based on an assurance case.

4.1 Goal Tree for AI System

To develop an assurance case for a system, we first set a top goal that the system needs to achieve in a specified context. Practical AI systems are expected to satisfy quality criteria so that users can use them in their office operations.

The ISO/ISE 9126 defines a quality model with six characteristics, namely functionality, reliability, usability, efficiency, maintainability, and portability [17]. In a project developing an MVP (Minimum Viable Product) of an AI system, we assess whether the AI system provides functions that meet the needs of users under specified conditions. Therefore, the top goal for developing AI systems as MVP is that the system satisfies functionality. The top goal is represented in Fig. 2.

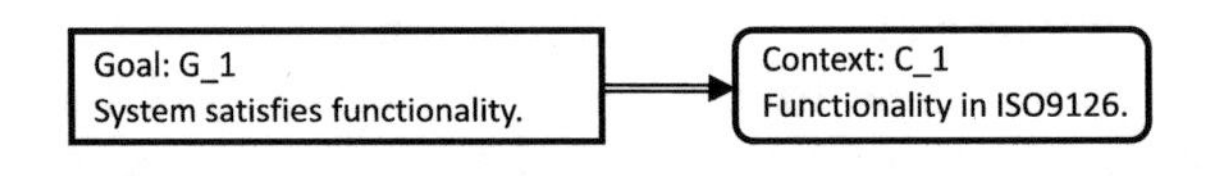

Fig. 2. Top goal

Functionality is further divided into sub-characteristics. These sub-characteristics are defined for general software systems [17] and they have been customized when applied to a specified application [18]. We extend functionality sub-characteristics for AI systems as shown in Table 1. Using these sub-characteristics, we decompose the top goal into sub-goals. We further decompose the sub-goals on suitability, accurateness, and interoperability. Finally, we obtain the goal tree for an AI system in Fig. 3 using these decomposed sub-goals.

4.2 Assurance Case and Project Assessment

Using the goal tree derived in the previous section, we assess the readiness for an AI system development project. When developing an assurance case, we assign evidence for each sub-goal. There are various artifacts such as discussion papers or meeting minutes developed through discussions between stakeholders before starting a project.

In the proposed method, we assign these artifacts as evidence of sub-goals (G5, G6, G7, ... G15). For example, we assume that there are the following discussions in a project developing an inquiry support system.

Table 1. Quality characteristics on functionality

Sub-characteristics	Functionality for software system	Functionality for AI system
Suitability	The system can perform the tasks required	The system contains optimal options for all possible inputs
Accurateness	The result is as expected	The system provides a correct option for input with a fixed probability
Interoperability	The system can interact with options systems	The system can interact with other systems (incl. human) that provide data for the system or use the system results
Security	The system prevents unauthorized access	Same as on the left
Compliance	The system is compliant with regulations	Same as on the left

- There are manual documents for a target operation and we develop answers for the inquiry service from these documents.
- Operators at a contact center refer to these documents during a phone conversation.
- The manual documents will be provided to the project team when starting a project.

These discussion outcomes described in discussion papers or meeting minutes can be considered evidence for G7. In this case, the assignment of evidence can be represented in Fig. 4(a).

However, there are some projects in which all items described in sub-goals are not discussed before starting the project. We assign a non-determined mark to

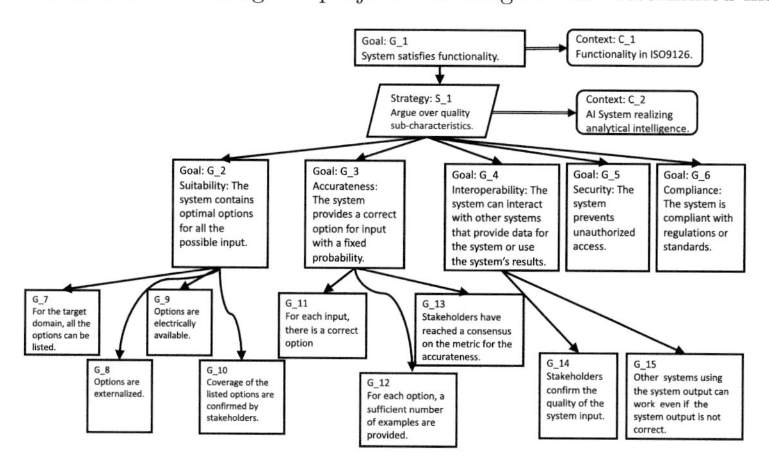

Fig. 3. Goal tree for AI system

such a sub-goal to explicitly show that there was no discussion on the sub-goal (Fig. 4(b)). There is a case in which developers confidently decide that some sub-goals do not need to be discussed before starting a project. For example, if there is a project in which a sub-goal is satisfied because of the AI engine specification, we introduce a new notation as shown in Fig. 4(c) to show this situation. Using this notation, we describe why evidence is not required for a sub-goal. If a sub-goal should be satisfied during a project, we can use this notation to track whether the sub-goal has been discussed.

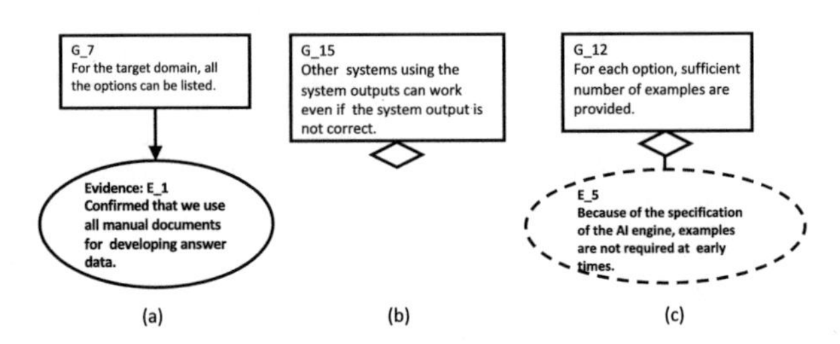

Fig. 4. Example of terminal subgoals ((a): with evidence, (b): without evidence, (c): for which evidence is not required)

When starting an AI system development project, we develop an assurance case based on the proposed method and assess how many sub-goals are satisfied. Figure 5 shows an example of the assurance case of a project developing an AI

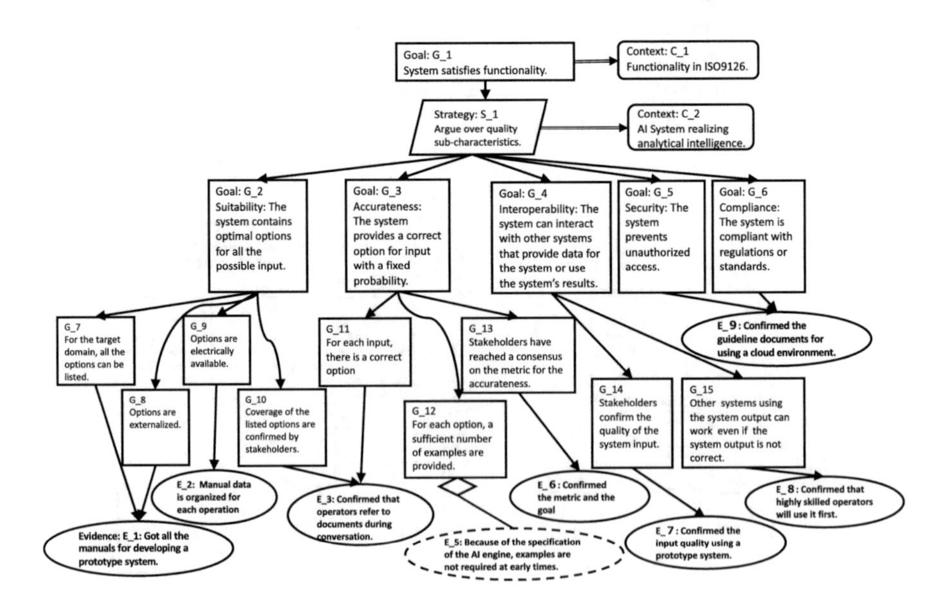

Fig. 5. Example of project readiness assessment

system supporting inquiry service. In this project, we developed the assurance case by referring to discussion papers and meeting minutes.

5 Experiment

We applied the proposed method to completed projects developing an AI system and assessed the maturity of the project preparation based on the discussion documents created before starting the project. Through this experiment, we derived sub-goals correlating to project success as project success factors.

In six projects, we developed AI systems for three types of business scenarios:

- Inquiry service support or automatic inquiry service
- Sales recommendation
- Screening service using documents

We divided these projects (A, B, ..., F) into three status groups based on the interviews with project managers. We categorized projects in which there were no serious issues during project delivery and subsequent projects had already started with status label 1. We categorize projects in which there were some issues due to mis-understandings between stakeholders about the system under consideration as status label 3.

For the projects, we developed assurance cases using the proposed method and assessed whether the sub-goals were satisfied. We assigned three labels (○, △, □) to 11 terminal sub-goals in the assurance case (G5, G6, ..., G15) as an assessment result. When a case in which all stakeholders confirmed that a sub-goal was satisfied, we assign a ○ label to the sub-goal. When a sub-goal was not discussed between stakeholders, we assigned a □ label to the sub-goal. We assigned a △ label to a sub-goal that stakeholders discussed but did not clearly confirm as fulfilled or when developers decided that the sub-goal need not be discussed. We showed sub-goal assessment results and project status score in Table 2.

Table 2. Sub-goal assessment results and project status

	G7	G8	G9	G10	G11	G12	G13	G14	G15	G5	G6	Status
Project A	○	○	○	○	○	△	○	○	○	○	○	3
Project B	○	○	△	○	□	○	○	○	△	○	○	3
Project C	○	○	△	△	○	△	○	□	△	△	△	2
Project D	○	○	○	□	○	△	○	△	△	□	□	2
Project E	□	□	□	□	□	□	□	□	□	○	○	1
Project F	○	□	□	△	△	□	○	○	□	○	○	1

From the assessment results in Table 2, we calculated the correlation scores (r) between the sub-goals and project status scores. In Table 3, we show sub-goals that are highly correlated to the project success score ($r \geq 0.75$).

Table 3. Correlation between sub-goals and project status scores

Sub-goal	G12	G15	G8	G9	G10
Correlation	0.891	0.891	0.866	0.750	0.750

From these correlation analysis results, we gathered the following successful factors for a project developing a system with analytical intelligence.

1. Sufficient number of training data (examples) for each option can be prepared (G12).
2. When a system has incorrect output, other systems (including users) can handle it practically (G15).
3. Data representing options for the system can be provided in a machine-readable format (G8, 9).
4. In a project, there is a consensus approach to confirming whether defined options will cover the target domain.

6 Discussion

By applying the proposed method to real projects, the success factors for executing practical AI projects have been identified. Based on the results of the experiments, we can predict whether we can successfully complete an AI system development project with the project assessment based on the assurance case. We can use our proposed method as a metric for AI project readiness.

In the first, third and fourth success factors, exhaustively defining options (possible outputs) for AI systems and preparing sufficient training examples are identified. These work items are generally important for machine learning systems and project members (both clients and vendors) should reach a consensus on the objectives and the significance of these work items. These are well-known factors for practitioners who understand AI technologies. On the other hand, the second success factor depends on how much experience the developers have with AI technology-related projects. This means that the second success factor may be neglected during project preparation. Our method identifies unknown success factors suitable for project assessment metrics.

A check list [19] can be used as a project readiness metric. However, the assessment items are listed as a flat format in a check-list. In contrast, our method represents them in a tree structure and relations between the assessment items can be considered. Therefore, when we focus on sub-characteristics from whole quality characteristics of functionality, we can prioritize sub-goals to be discussed by the developed assurance cases.

By applying our method to several project, we can develop a model for predicting project risks quantitatively. To develop a prediction model, we need to calculate a score for a developed assurance case based on how many sub-goals have evidence. For this, attribute GSN [20] can be one of the approaches considered. Also, we need to introduce a monitoring mechanism into our method to update undetermined sub-goals within suitable timing. These are our future works.

7 Conclusion

In this research, we considered AI System development projects that include machine learning techniques for office environments. For such AI system development projects, we proposed a method that uses an assurance case to determine project success factors and assess project readiness.

To apply AI technologies in a project, clients' project members need to prepare training data and determine how to handle the application's output if the machine learning output is incorrect. Therefore, project members (both clients and vendors) should have the same understanding of the objective and the significance of each work item assigned to them. However, this depends on the project leader's skill and experience.

To solve this issue, we proposed a method to assess an AI system development project by using an assurance case based on quality sub-characteristics on functionality. We applied the proposed method to real projects and could derive project success factors.

References

1. Berntsson-Svensson, R., Aurum, A.: Successful software project and products: an empirical investigation. In: Proceedings of the 26th ACM/IEEE International Symposium on Empirical Software Engineering, pp. 144–153 (2006)
2. Quichiz, L.P., Oré, S.B.: It demand management in organizations: a review. In: Proceedings of the 18th International Conference on Computer Modeling and Simulation, pp. 95–99 (2017)
3. Chen, H.M., Kazman, R., Haziyev, S.: Strategic prototyping for developing big data systems. IEEE Softw. **33**(2), 36–43 (2016)
4. Kim, M., Zimmermann, T., DeLine, R., Begel, A.: The emerging role of data scientists on software development teams. In: Proceedings of the 38th International Conference on Software Engineering, pp. 96–107 (2016)
5. Song, I.Y., Zhu, Y.: Big data and data science: what should we teach? Expert Syst. **33**(4), 364–373 (2016)
6. Wang, X., Mylopoulos, J., Guizzardi, G.: How software changes the world: the role of assumptions. In: Proceedings of the 10th IEEE International Conference on Research Challenges in Information Science, pp. 1–12 (2016)
7. Kelly, T., McDermid, J.A.: Safety case construction and reuse using patterns. In: Proceedings of the 16th International Conference on Computer Safety, Reliability and Security, pp. 55–69 (1997)
8. Kelly, T., Weaver, R.: The goal structuring notation – a safety argument notation. In: Proceedings of Dependable Systems and Networks Workshop on Assurance Cases (2004)
9. Yamamoto, S., Matsuno, Y.: An evaluation of argument patterns to reduce pitfalls of applying assurance case. In: Proceedings of the 1st International Workshop on Assurance Cases for Software-Intensive Systems (ASSURE), pp. 12–17 (2013)

10. Doss, O., Kelly, T.: Addressing the 4+1 software safety assurance principles within scrum. In: Proceedings of the Scientific Workshop Proceedings of XP2016, pp. 17:1–17:5 (2016)
11. Bloomfield, R., Bishop, P., Butler, E., Netkachova, K.: Using an assurance case framework to develop security strategy and policies. In: Proceedings of the International Conference on Computer Safety, Reliability, and Security, pp. 27–38 (2017)
12. Sklyar, V., Kharchenko, V.: Challenges in assurance case application for industrial IoT. In: Proceedings of the 9th IEEE International Conference on Intelligent Data Acquisition and Advanced Computing Systems, pp. 736–739 (2017)
13. Boness, K., Finkelstein, A., Harrison, R.: A method for assessing confidence in requirements analysis. Inf. Softw. Technol. **53**(10), 1084–1096 (2011)
14. Kaneko, T., Yamamoto, S., Tanaka, H.: CC-case based on system development life-cycle process. In: Proceedings of the International Conference on Computer Security and Digital Investigation (COMSEC), pp. 29–35 (2014)
15. Varkoi, T., Nevalainen, R., Mäkinen, T.: Process assessment in a safety domain. In: Proceedings of the 10th International Conference on the Quality of Information and Communications Technolgy, pp. 52–58 (2016)
16. Sternberg, R.J.: Successful Intelligence: How Practical and Creative Intelligence Determines Success in Life. Simon & Schuster (1996)
17. ISO/IEC TR 9126: Software engineering - product quality (2000)
18. Djouab, R., Bari, M.: An ISO 9126 based quality model for the e-learning systems. Int. J. Inf. Educ. Technol. **6**(5), 370–375 (2016)
19. Yamamoto, S.: A continuous approach to improve it management. In: Proceedings of the International Conference on Enterprise Information Systems (CENTERIS), pp. 27–35 (2017)
20. Yamamoto, S.: Assuring security through attribute GSN. In: Proceedings of the 5th International Conference on IT Convergence and Security (CIITCS), pp. 1–5 (2015)

Estimation of Business Rules Using Associations Analysis

Takuya Saruwatari(✉), Akio Jin, Daisuk Hamuro, and Takashi Hoshino

NTT Software Innovation Center, Nippon Telegraph and Telephone Corporation, 2-13-34 Kounan, Minato-ku, Tokyo 108-0075, Japan
takuya.saruwatari.gz@hco.ntt.co.jp

Abstract. Recently, IT system renewal projects are increased along with aging of IT systems in long term operation. In such a project, it is necessary to extract the business rules that were implemented in legacy system. In this paper, a method is proposed that estimate business rules from data that stored in the legacy system. An association analysis is used in the proposed method. The proposed method is evaluated using pseudo data.

Keywords: Legacy system · System renewal · As-Is system analysis Business rules · Data mining

1 Introduction

Recently, IT system renewal projects are increased along with aging of IT systems in long term operation. In such a project, a legacy system is replaced by new system. There are many cases that legacy system that is target of IT system renewal project was used for long term and is huge systems. In the IT system renewal projects, it is necessary to extract the business rules that were implemented in legacy system. But, in the legacy system that has been in operation for a long time, it is difficult to grasp these business rules. There are the following factors that makes it difficult to grasp these business rules. "Business rules were not documented." "Change of business rules accompanying additional system development were not documented."

In this paper, a method that estimate business rules from legacy system is proposed. The proposed method can be used in IT system renewal projects. In the proposed method, stored data in legacy system are used as input data and association analysis is used to estimate business rules. An association analysis is one of the data mining techniques. Estimated rules by the proposed method can be used as candidate to grasp correct business rules.

The remainder of this paper is organized as follows. Section 2 describes key related researches. Section 3 details the proposed method. In Sect. 4, we evaluate our method through experiment with pseudo data. Section 5 have several discussion that related our proposed method. Section 6 present our conclusions.

M. Virvou et al. (Eds.): JCKBSE 2018, SIST 108, pp. 33–42, 2019.
https://doi.org/10.1007/978-3-319-97679-2_4

2 Related Works

There are some techniques that analyze software source code or execution trace to estimate business rules implemented in legacy systems [1, 2]. These techniques are called reverse engineering. In common reverse engineering techniques, business rules are estimated from software source code, execution trace or etc. In the proposed method, stored data in database of legacy system are used to estimate the business rules. Therefore, it is expected that business rules that cannot be estimated by common reverse engineering technique can be estimated using the proposed method. That is, we think that two techniques can be used complementary.

Association analysis is one of the data mining techniques [3]. In this technique, association rules are extracted from transaction data that is stored in database. Extracted association rules are used as new knowledge. In order to perform association analysis on a computer, algorithms such as Apriori [4] have been developed. Data mining techniques had developed as information analysis techniques for marketing area. For that reason, association analysis is mainly used for analyzing customer purchasing trends in marketing. In the proposed method, there is a unique point that association analysis is used to estimate the business rules that are implemented in legacy systems. Moreover, typically, numerous rules are extracted by association analysis. In the proposed method, the technique that summarize multiple rules is proposed. Concept of "Same meaning rules" are introduced in Sect. 3.1. "Same meaning rules" is related to the summarizing technique.

3 The Proposed Method

In this paper, the method that estimate business rules from stored data in databases is proposed. Association analysis is used in the proposed method. It is one of the data mining techniques. In this section, business rules, input and output data are explained. They are treated in the proposed method. A process of the proposed method are also shown.

3.1 Business Rules

There are various kind of business rules. One of the definition of business rules is as follows.

"A business rule is a statement that defines or constrains some aspect of the business" [5].

Although there are many kind of business rules, business rules that are treated in this paper are restricted. Type of business rules that can be expressed as "If A then B" are treated in this paper. Example of this type of business rules are shown as follows. "Customer service application rules that are decided by contract status or customer age", "Approval rules of organization that are depend on project type and size of amount of money".

Note that, even if "If A then B" is correct, "If B then A" is not necessarily correct. 2 rules that is shown above are distinguished.

Same Meaning Rules

In this paper, rule collection satisfying the following conditions are defined as "Same meaning rules".

- The conclusion parts of rules are same.
- The condition part of any rules are included the condition part of one rule of the collection.

For example, next two rules are "Same meaning rules". Rule.1: $A \wedge C \rightarrow B$. Rule.2: $A \rightarrow B$. Both of rules have same conclusion part. That is, conclusion part of both of rules are "B". And, condition part of Rule.1 is included by condition part of Rule.2. In this case, if Rule.2 is correct, Rule.1 is always correct.

Moreover, "Primary rule" is defined. "Primary rule" is most simple rule of "Same meaning rules". In the above example, Rule.2 is defined as "Primary rule".

3.2 Input Data of the Proposed Method

Input data of the proposed method is the data that is stored in database. In the proposed method, one table data of database is used in one analysis. If there are some tables that is candidate of analysis, it is necessary to analyze each table separately.

3.3 Output Data of the Proposed Method

In the proposed method, the business rules that can be defined as "If A then B ($A \rightarrow B$)" are treated. Therefore, it is necessary to know the combination of the condition part (A) and the conclusion part (B) to grasp the business rules which analyst want to know. In the proposed method, the list of combination of A and B is outputted. Furthermore, labels that indicate "Same meaning rules" are putted to each rules and labels that indicate "Primary rule" are putted to primary rule.

3.4 Procedure of the Proposed Method

The proposed method has 3 steps to estimate rules. Each steps are described below.

- Step 1: Preprocessing
  Target columns of analysis in the input data are selected. Analyst can be selected columns that (s)he want to analyze. The columns that are not selected are ignored of the analysis. Next, transaction data are created from input data. One transaction record contains data which are contained in one record of input data. Here, transaction record has the same meanings as basket (as data mining term). At creating the transaction data in Step 1, the column name label of the data is not used.
- Step. 2: Association analysis
  Association analysis is performed to transaction data that is obtained in Step 1. In this paper, "apriori function" that is included in R [6] is used. "Apriori function" is implementation of Apriori [4] on the R. When performing association analysis, the following parameter conditions are used. Support value of rules is 0.01 or more. Confidence value of rules is 1. Here, support and confidence value are attributes of

rules in association analysis. By setting the confidence value to 1, it is possible to obtain only the rule that the conclusion part is satisfied when the condition part is satisfied. That is, by setting the confidence to 1, only the rule that can be presented as "If A then B" can be obtained.

- Step 3: Identification of "same meaning rules"
  "Same meaning rules" are identified in the obtained rules at Step 2. Then, labels that present "Same meaning rules" are given to each rules. And, each primary rules are identified in each "Same meaning rules". Primary rule is most simple rule of each "Same meaning rules".

4 Evaluation Experiment

In this paper, proposed method is evaluated using pseudo data that is made for evaluation experiment. Several rules were embedded in the pseudo data. In this section, content of pseudo data and results of evaluation experiment are described.

4.1 Pseudo Data for Evaluation Experiment

Data that several rules were embedded was prepared as pseudo data for evaluation experiment for the proposed method. The prepared pseudo data has several columns. These columns are shown in Table 1. The prepared pseudo data has 1,000 record in it. Contained data of each columns are also shown in Table 1. 5 kind of data are contained in column A, B, E, and F, and 2 kind of data are contained in column C and D, as shown in Table 1.

Table 1. Columns of pseudo data for evaluation experiment.

Column name	Contained data
A	A1, A2, A3, A4, A5
B	B1, B2, B3, B4, B5
C	C1, C2
D	D1, D2
E	E1, E2, E3, E4, E5
F	F1, F2, F3, F4, F5

11 rules are embedded in the pseudo data. Embedded rules inside of the pseudo data are shown in Table 2. The rules are expressed using logical expression. R01 to R10 show that the data of column A and column B are synchronized. R11 shows the relationship in which the data of column E is determined from the data of column C and column D. Data that is not decided by embedded rules in the pseudo data are set to random value in the range that is described in Table 1.

Table 2. Rules embedded in the pseudo data

Rule ID	Embedded rules
R01	A1 → B1
R02	B1 → A1
R03	A2 → B2
R04	B2 → A2
R05	A3 → B3
R06	B3 → A3
R07	A4 → B4
R08	B4 → A4
R09	A5 → B5
R10	B5 → A5
R11	C1∧D1 → E1

Part of pseudo data is shown in Table 3 as sample. In this sample, two records are shown.

Table 3. Sample of pseudo data

A	B	C	D	E	F
A1	B1	C2	D2	E1	F1
A1	B1	C2	D1	E5	F4

4.2 Evaluation Experiment Results

In this section, evaluation experiment results are described. In the evaluation experiment, 1,105 rules were obtained by execution of Step 2. In Step 3, the label to identify "Same meaning rules" is given to rules that are obtain in Step 2. As a result, 29 kind of "Same meaning rules" were obtained. That is, 1,105 rules that were obtained in Step 2 were classified into 29 kind of collections. Most simple rules (rule that was given "primary rule" label) of each collection are shown in Table 4.

Table 4. Rules that are given "primary rule" label.

No.	Rule head	Rule body	Label
1	C1, D1	E1	1
2	B5	A5	2
3	A5	B5	3
4	B1	A1	4
5	A1	B1	5
6	A3	B3	6
7	B3	A3	7
8	A2	B2	8
9	B2	A2	9
10	B4	A4	10
11	A4	B4	11

(*continued*)

Table 4. (*continued*)

No.	Rule head	Rule body	Label
12	D1, E3	C2	12
13	D1, E4	C2	13
14	D1, E2	C2	14
15	D1, E5	C2	15
16	A3, E4, F3	C2	16
17	B3, E4, F3	C2	17
18	C1, E3	D2	18
19	C1, E5	D2	19
20	C1, E4	D2	20
21	C1, E2	D2	21
22	B5, E1, F5	C1	22
23	A5, E1, F5	C1	23
24	B5, D1, E1, F4	C1	24
25	A5, D1, E1, F4	C1	25
26	B1, E1, F5	D1	26
27	A1, E1, F5	D1	27
28	B4, C1, E1, F5	D1	28
29	A4, C1, E1, F5	D1	29

By seeing from No.1 to No.11 in Table 4, it can be understood that R_11 (No.1), R_10 (No.2), R_09 (No.3), R_02 (No.4), R_01 (No.5), R_05 (No.6), R_06 (No.7), R_03 (No.8), R_04 (No.9), R_08 (No.10) and R_07 (No.11) which are shown in Table 2 could be estimated. That is, all rules which were embedded in pseudo data could be estimated. From these results, precision and recall value are calculated. They are shown below.

Precision = 0.38 Recall = 1

Table 5. Frequency distribution for each support value of the obtained rules

Range (support)	Frequency
0.00–0.02	10
0.02–0.04	0
0.04–0.06	8
0.06–0.08	0
0.08–0.10	0
0.10–0.12	0
0.12–0.14	0
0.14–0.16	0
0.16–0.18	0
0.18–0.20	4
0.20–0.22	6
0.22–0.24	1
Total	29

Frequency distribution for each support value of the obtained rules that has "primary label" is shown in Table 5 and Fig. 1. Minimum value of support value range are shown at x axis of Fig. 1. It is understood that estimated primary rules are categorized 2 groups from Fig. 1. Rules which support value is 0.18 or more are one group and rules which support value is less than 0.06 are another group.

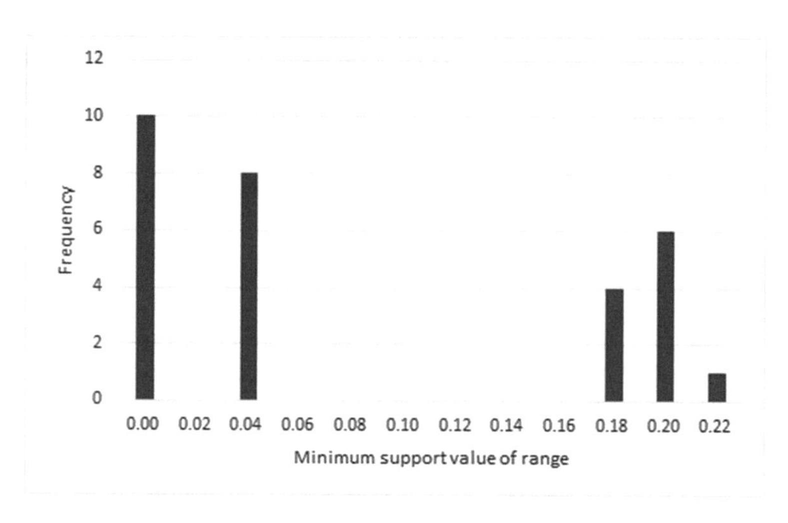

Fig. 1. Frequency distribution for each support value of the obtained rules

5 Discussions

5.1 Usefulness of Proposed Method

Business rules that were shown in Table 4 could be estimated using proposed method. Moreover, estimated rules included all rules that were embedded to pseudo data beforehand (Recall value = 1). That is, it was shown that the proposed method is usefulness to estimate business rules from stored data.

The purpose of the proposed method is to present candidate rules to analysts. Considering that purpose, it can be said that Precision value (0.38) is also not low value.

5.2 Reducing the Number of Estimated Rules

The purpose of the proposed method is to present candidate rules to analysts. In the proposed method, association analysis is used to estimate candidate rules. When estimated rules are shown to rule analyst, it is better to narrow down to those that are likely correct rules. But, just by using association analysis, huge number of rules are estimated. In the experiment of this paper, 1,105 rules were estimated by Step 2. In such a situation, it may be difficult for rule analysts to pick out correct rules from estimated rules, because, huge number of rules are presented to analyst. Therefore, in the proposed method, Step 3 are performed after association analysis. In the Step3, "same meaning rules" label are putted to each rules that were estimated in association analysis in Step 2. Rules that has same label are same kind of rule. That is, same meaning rules. In the experiment, 1,105 rules that were estimated by association analysis in Step 2 were categorized to 29 kind of "same meaning rules". Moreover, most simple rule in each "same meaning rules" are identified as "primary rule". By presenting the primary rules to the analyst, it was able to drastically reduce the rules which were proposed to the analyst. In the evaluation experiment, presented rules were reduced to about 0.026 times by Step 3. It is expected that burden of rule analyst can be reduced by performing Step 3. Step 3 is strong point of the proposed method compared with simple application of association analysis.

5.3 Frequency Distribution for Each Support Value of the Obtained Rules

Assuming that rules that are labeled "primary label" are estimated rules by proposed method. Under such an assuming, estimated rules can be divided into two groups with a degree of support value less than 0.06 and 0.18 or more (see Fig. 1). Rules that the support value ranged from 0.06 to 0.18 were not estimated. Moreover, all support values of rules that are embedded before experiment were included between 0.06 and 0.18. That is, it turns out that the business rules that were estimated correctly and the business rule that were not estimated correctly divided relatively clearly in the experiment. In the experiment of this paper, it was found that rules these support value were 0.18 or higher should be focused on. But, it is necessary to pay attention that a certain number of data that were followed correct rules were contained in this experiment. Therefore, there was a possibility that the correct rule could be separated relatively clearly. Consideration must also be given to the fact that the rule whose support value is 0.04 is a rule that appears in 4% of the entire data. Since, in this experiment, 1,000 transactions were in the data, 40 transactions were followed the rules whose support value was 0.04. Since this evaluation, it turned out that it may be useful to use support value for selecting rules. The study of how to utilize the support value is a future work.

5.4 Same Data in Different Columns

In the common association analysis, column names are not used. Therefore, when same data is included in multiple columns, adverse effects may occur. For example, in the pseudo data used in the evaluation experiment, if "A1" or "C1" was included column B, there is a possibility that incorrect rules are estimated by effect of these data. In order to prevent such an adverse effect, it is necessary to prevent becoming the same data label, when data are in different columns in Step 1. For example, assigning the column name to each data as prefix, and the like are conceivable.

5.5 Effect of Fault Data

If fault data that does not conform the rules is contained in the input data, it is conceivable that correct rules cannot be estimated.

In the proposed method, only the rules whose confidence value (of association analysis) is 1 is assumed as the estimated rules. By setting the confidence value to 1, it is possible to obtain only the rules that the conclusion part is always satisfied when the condition part is satisfied. If fault data that match the condition part of the rule but do not match the conclusion part of the rule is exist, that rule cannot be estimated. Because, confidence value of the rule is not become 1. For example, in the evaluation experiment in the paper, if transaction that column C value is "C1", column D value is "D1" and column E value is "E2" was exist, R11 ($C1 \wedge D1 \rightarrow E1$) cannot be estimated. That is, fault data causes a situation where correct rules cannot be estimated.

About this issue, the following countermeasures can be considered. Before application of Step 1, performing data cleansing to remove fault data. The following means are also conceivable. Add the rules that confidence value not equal 1 to the estimation result. However, if adding the rules that confidence value not equal 1 to the estimation result, it is conceivable that the precision value will decrease. It is necessary to consider how much confidence value and precision value are acceptable. This is future work.

5.6 Limit

In this paper, there are limits in the following points.

- The pseudo data that was used in evaluation experiment in this paper is very simple. There is a limit that there are few type of pseudo data that was used in the experiment.
- The proposed method was not applied to real situation, yet. There is a limit that evaluation experiment in this paper was only using pseudo data. In the future, we should perform evaluation experiment with concrete data.

6 Conclusion

Recently, systems have been used long time are increasing. These systems are called legacy systems. IT system renewal projects are increased along such a situation. In IT system renewal project, the legacy system is replaced with a new system. In such a project, it is necessary that business rules which are implemented in legacy system must be grasped. In this paper, to respond to this request, business rule estimation method is proposed. The proposed method has 3 steps. Association analysis is used in step 2. Input data of the proposed method is stored data in legacy system. Evaluation experiment to confirm effectiveness of the proposed method are also performed. In the evaluation experiment, pseudo data was used. As a result of the evaluation experiment, it is shown that business rules can be estimated using the proposed method. That is, effectiveness of the proposed method was shown.

The future works are evaluation of proposed method through real situation, comparison with other algorithm and blush up of the proposed method etc.

References

1. Acharya, M., Xie, T., Pei, J., Xu, J.: Mining API patterns as partial orders from source code: from usage scenarios to specifications. In: Proceedings of the 6th Joint Meeting of the European Software Engineering Conference and the ACM SIGSOFT Symposium on the Foundations of Software Engineering, pp. 25–34. ACM (2007)
2. Gabel, M., Su, Z.: Javert: fully automatic mining of general temporal properties from dynamic traces. In: Proceedings of the 16th ACM SIGSOFT International Symposium on Foundations of Software Engineering, pp. 339–349. ACM (2008)
3. Agrawal, R., Imieliński, T., Swami, A.: Mining association rules between sets of items in large databases. ACM SIGMOD Rec. **22**(2), 207–216 (1993)
4. Agrawal, R., Srikant, R.: Fast algorithms for mining association rules. In: Proceedings of the 20th International Conference on Very Large Data Bases, pp. 487–499 (1994)
5. Hay, D., Healy, K.A., Hall, J.: Defining business rules-what are they really. The Business Rules Group, vol. 400 (2000)
6. The R Project for Statistical Computing. https://www.r-project.org/. Accessed 7 Mar 2018

Moral Education for Adults for Information Ethics to Effect the Unknown Problem

Keiichiro Abe[1(✉)] and Takako Nakatani[2]

[1] Higashichikushi Junior College, 5-1-1 Shimoitozu, Kokurakita-ku, Fukuoka, Japan
abe@hcc.ac.jp
[2] The Open University of Japan, 2-11 Wakaba, Mihama-ku, Chiba, Japan

Abstract. We develop an educational method for adults. The method helps adults think how they should behave in a society. They are also expected to deal with information according to the so-called "information ethics." In order to educate their information ethics, our method of moral education is applied. We evaluate the method whether the moral education contributes to the development of adults' information ethics or not. We apply a self-debate to the moral education. The self-debate is thinking oneself, worrying this and that and putting together oneself. According to case studies, we could clarify that subjects were able to reflect upon themselves and acquire knowledge about information ethics through the self-debate. In this paper, we report the result of the education.

Keywords: Information ethics · Moral · Dilemma · Self-debate

1 Introduction

An information moral education is an education on information ethics required for all the people in the information society [1]. Information moral is a part of our daily life moral [2]. The daily life moral is included in the curriculum of compulsory education. We assume that "Moral Education for Adults" (it carries out the following "MEA") helps the adults understand the information ethics deeply than now.

There are various researches on the education of information ethics for adults. For example, Okada reported case studies of which subjects were university students [3]. However, our ethical knowledge does not work for an unknown information ethics well. Moreover, even if we encounter the problem of known information ethics, we may not act ethically without morality. In order to educate information ethics more effectively, we should educate morality to adults. The moral education in Japan is included in the curriculum of compulsory education, "special subject morality [4] ". All the adults have studied morality. This is the reason why there are few study cases on MEA. In this paper, we rethink about MEA and develop a method of the education.

This paper is outlined as follows. We discuss related work in Sect. 2. Section 3 describes the approach of our method. Section 4 describes a case study. In Sect. 5, we discuss the results and in the final section, we conclude this paper.

M. Virvou et al. (Eds.): JCKBSE 2018, SIST 108, pp. 43–52, 2019.
https://doi.org/10.1007/978-3-319-97679-2_5

2 Related Work

(A) L. Kohlberg's Moral dilemma

L. Kohlberg introduced six stages of moral judgment through practical research on moral education. The moral dilemma is the moral conflict concerning the activities that are selected or not selected on the basis of moral value. As each choice has moral merits and demerits, students cannot justify each choice immediately. In moral dilemma education, a teacher provides a hypothetical story and a two-alternative question for students who take the course. After the students were conflicted about the value of morality, they choose one of a two-alternative questions and they write the reason for their choice. Through the moral discussion, each student rethinks their original opinion. Finally, the students select the specific moral value and they report their opinions to the teacher.

In Table 1, the stage of development of L. Kohlberg's morality [1] is shown. The six stages are categorized into three levels of two stages each. Morality is developed from the first to the sixth stage.

Table 1. Six stages of moral judgment by L. Kohlberg [1]

Level	Stage
Level 1: Preconventional	Stage 1. Heteronomous morality
	Stage 2. Individualism, instrumental purpose, and exchange
Level 2: Conventional	Stage 3. Mutual interpersonal expectations, relationships, and interpersonal conformity
	Stage 4. Social system and conscience
Level 3: Postconventional or principled	Stage 5. Social contract or utility and individual rights
	Stage 6. Universal ethical principles

Their educational method may be effective in children. Our target subjects are adults. In our previous work, we clarified that debate the question of the moral dilemma with group members does not work for adults. We need another way for the education of adults.

(B) Research on the learning materials for information ethical dilemma of students

A learning material for information ethical dilemma is one of the materials for the moral dilemma. Murakami et al. validated the materials of a class on information ethics [6]. The participants included approximately 2/3 incoming students at Hiroshima University. Five original problems were provided in the class. Each problem described one situation in 500 to 700 characters in Japanese and had two options of approximately 200 characters in Japanese each. Every student refuted the opinion, which he/she did not choose from the viewpoint of information ethics.

Inagaki et al. proposed the rate of two choices for moral dilemma [7]. The five problems were evaluated by the rate of each choice. It was discussed from the rate of each choice whether the student felt a dilemma with regard to the five problems. If one of the options was supported by at least one-third of the students, they recommended that the problem could be accepted as a dilemma problem. Furthermore, according to the analysis based on the theory of the developmental stages of L. Kohlberg's morality [1], they expected college students to have grown from stage 4 to stage 5. One selection of each question indicated the fourth stage of the development of morality, whereas the other selection indicated the fifth stage. Therefore, the students who chose the latter were regarded in the higher level of the morality than the students who chose the former.

Those researchers regarded that the morality was developed by refuting documents. However, in order to invite students to the higher stages of morality, we have to develop a mechanism of evaluation of subject's stage.

3 Research Approach

In this research, we developed MEA and evaluate the efficiency of extending an understanding of information ethics. We consider whether moral education will be an effective method to help adults understand the information ethics. The Japanese government announced the following guidelines for teaching "special subject morality [8]." "A moral education which gazes at self-based on an understanding of moral value needs to think the based on information ethics deeply."

The basis of information ethics is the sympathy of others, the laws, and the rules. However, this Japanese government announced the following guidelines is for school children. If we educate adults to raise morality, even if they will encounter a strange phenomenon, they may be possible to work through the problem of information ethics. The hypothesis of this research shall be as follows. "MEA is effective in order to deepen an understanding of the information ethics." In order to verify this hypothesis, we develop the method of MEA and show clearly whether MEA contributes to an understanding of the information ethics. The procedure of verification is as follows.

1. We explain measuring the educational result of information ethics to students. And students solve the information ethics problems.
2. L. Kohlberg [9] insisted "We respect children's autonomy." A teacher should not force and/or control students' ability to think. However, the discussion of the development of moral education is out of the scope of their paper. But we only refer to the developmental stage of morality. Since the self-debate is an internal argument, students cannot feel compelled to think. The result of the self-debate shows how many moralities the student raised.
3. The teacher evaluates each student's development stage of L. Kohlberg's morality [5] by a document was written by each student.
4. Students solve the same information ethics problems as 1. If they have a higher score of the information ethics problems than the before, MEA effects the information ethics education.

In this research, we think the students' morality increases by self-debate.

3.1 Examination of MEA - Evaluation of the Validity of Discussion -

In order to investigate why research of MEA is not done, we tried the technique of L. Kohlberg [5] to the adult as one of the moral education. It is because the reason for having chosen the technique of him can evaluate morality nature, so it is easy to compare the educational result. Moreover, this is because it is easy to make the teaching materials tailored for the target. "Heinz's dilemma [5]" of L. Kohlberg's representation subject was tried to the adult on December 16, 2016. The subject created for junior and senior high school students [10, 11] was tried to the adult on January 13, 2017.

As a result, discussion needed for moral education finished immediately. Students had not the time to improve morality through arguments. It seems that we devised so that they might debate, but they do not debate in order to make a judgment on subjects as an already-known fact. In order for us to find the moral education technique for the adults, it is close to discussion for not discussing, or the effective method better than discussion.

3.2 Examination of MEA - Evaluation of the Self-debate -

One of the methods of carrying out moral education, without discussing, has "sympathy as morality facts [12]." This is regarding those who are troubled with a pain as one's pain. Also in the morality performed by compulsory education, feeling pain and the pain of the characters of a book is performed as the first step of moral education in Japan. If students can feel the opinion as their pain through being greatly troubled by "moral dilemma" education, we will stand to the starting line of MEA.

In this research, the "self-debate system" is defined and we study whether morality is improved. Suppose that students refute the opinion which protects another side which they did not choose in the moral dilemma. There is also research to see the result of debates by the lesson of morality [13]. However, the self-debate is excellent in the following three points compared with "a discussion and a debate (it is called the following "2 Technique")."

(1) The students can do an argument to high morality.
(2) The students can certainly speak the opinion of them.
(3) Unnecessary information to moral education is pushed out.

The reason of (1) is that the large student of tells lower morality and a risk of all the groups becoming the argument on lower morality is because it is small compared with 2 Techniques. The reason of (2) is that it cannot perform by 2 Technique that all the students say their opinion equally. If the teacher can control a class to the self-debate since conversation is unnecessary, but the reason of (3) does not examine the contents of the talk by 2 Technique, the teacher cannot stop the conversation. It is because a risk of unnecessary information getting across to students is high.

Therefore, just the self-debate is suitable to MEA.

4 Case Study

4.1 Overview

This research targeted three groups of Higashichikushi Junior College. The 1st is 21-second graders (it carries out the following "the 2017 students") who graduated in 2017. The 2nd will be 26-second graders (it carries out the following "the 2018 students") of the graduation hope in 2018. The 3rd is 27 first graders (it carries out the following "the 2019 students") who are the 2019 graduation schedules. The questionnaire of Information ethics did to 2019 student before-and-after the morality lecture and did to 2018 student before-and-after the general lecture in November 2017.

In this research, the 2019 students were divided into two groups: one is the "Morality G" and the other is the "General G," and they attended the morality lecture and the general lecture, respectively. Through the education, two groups were isolated, so that we could compare the results of their education. The results were measured by the rate of correct answers of the examination on information ethics. We controlled both groups not to exchange the information about their educational contents. In the case study, we provided one personal computer to each student, so that they could use LMS [14] and report their answers via the LMS by themselves.

4.2 The Flow of MEA

The educational process for Morality G is as follows.

(1) Before the lecture, we carried out the 10 questionnaires of Information ethics to the student.
(2) The students read the first subject of teaching materials. The subject presented two selective actions of which moral values were different. Morality G reported their selections of the alternatives and the reasons for their selections.
(3) We gave the lecture on the development of morality.
(4) We check the changing of the moral development of students. The students read the second subject of the teaching materials. The second subject was different from the first subject since the second one included supporting opinions for both of the selections. The students selected one of the alternatives and they reported their selections with their counterarguments against the supporting opinion of a selection that they did not select.
(5) After the lecture, the student carried out what changed only orders of questions that used in the process (1).

Notes: General G attended a general lecture only instead of (2)–(4).

4.3 The Questionnaire of Information Ethics Education

Examination questions were extracted from the drill of information ethics [15] with some modifications. They were the 10 either-or questions for measuring the knowledge of students' information ethics. Table 2 shows a part of the questionnaire.

Table 2. A part of the questionnaire

Q7. The videotape which had been doubled a TV program was sold to a friend. Choose an answer suitable for this situation
A. It is an illegal sale, even though its price is low enough
B. You can sell, if its price is as low as the price of the tape
Q10. You received an advertising e-mail with your cellular phone. The email mentioned that you can cancel the receipt of such emails by sending an empty email to a certain address, if you do not want to receive these emails from the sender. If you do not want to receive the emails any more, choose an answer suitable for this situation
A. You delete the email as soon as possible, and you will never send any emails to the designated address
B. You will send an e-mail for terminating the receipts, if and only if. you have confirmed that sender is trustworthy

We evaluate the result of the examination before-and-after the lectures. Though both of the questionnaires were composed with the same contents, the turns of the questions were modified. Because we had to mitigate a threat of students' selection with their memories.

4.4 The Evaluation Method of MEA

First of all, before-after the lectures, we marked their answers and opinions on the basis of the stages of moral judgment by Kohlberg [5]. The results would be used as the pre-examination's results.

4.5 Results

Table 3 summarizes the result of the first subject. The number of choosers and their rate are shown in the left column and the averages of the developmental stage of the morality of students group are shown in the right column. We measured the developmental stage of the morality of students through their opinions with the standard of Kohlberg [5]. According to the results, we could conclude that the developmental stage of the morality of the students is beyond 4. Further, their stages are independent of their choices.

Table 3. The result of the subject 1.

Selection	The number of selectors, and the rate of each selection	Average of the developmental stage of morality measured by the students' reports of their opinions
A	2017 students: 11 (68.8%)	2017 students: 4.1
	2018 students: IS (69.2%)	2018 students: 4.3
B	2017 students: 5 (31.3%)	2017 students: 4.0
	2018 students: 8 (30.8%)	2018 students: 4.0

The average of the developmental stage of the morality of both of the 2017 students and the 2018 students was approximately four. Their average value fulfilled the stage expected as that of college students. Since both of the answers marked A and B were supported by over 30% students, this teaching material of dilemma of moral fulfilled the constraint proposed by Murakami et al. The 2017 students and 2018 students expressed many opinions for the answer of A. We could get the same results from the 2019 students. The rate of answers was almost the same.

Figure 1 represents the results of the development stage of the morality observed in the "moral dilemma" education for the 2019 students. According to the Fig. 1, a half of the students in the stage 4 could develop their stage to 5. Similarly, the rest of students in the stage 3 could develop their stage to 4. We could conclude that our education of morality with moral dilemma had been successfully performed. However, there were students who could not develop their stage to 5.

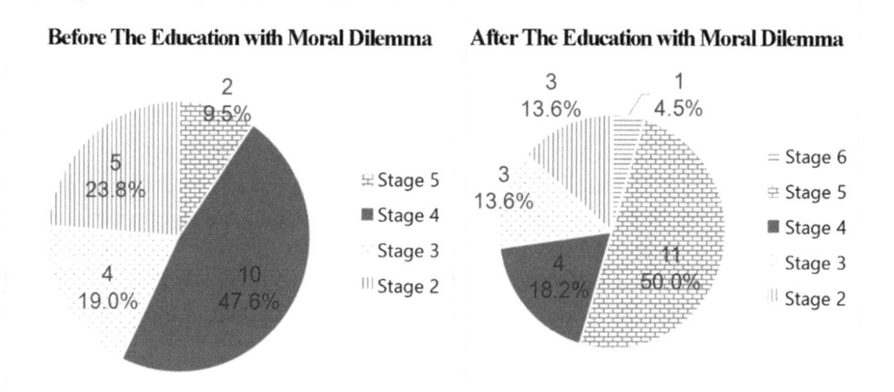

Fig. 1. The developmental stage of the morality of the 2019 students.

Though there was no student in the highest stage 6 before the education, a student could be developed her stage to the highest stage after the moral dilemma education. Before the education, the rate of students in the lower stages than the expected stage, 4, was 42.8%. After the education, the rate was decreased to 27.2%.

Figure 2 represents the result of the examination of information ethics of pre- and post-lecture.

In Fig. 2, the marks of plus and minus represent the difference between the result of the pre-lecture and that of the post-lecture. For example, "+3" means that the student increased her score for 3 points. The students in the General G were the 2018 students, and the students in the Morality G were the 2019 students. Since the size of samples was small, we could not apply statistical analysis to the samples. We compared the number of correct answers of pre- and post-lectures and analyzed the contents of the answers.

Interestingly, the students who had got worse from the examination of the pre-lecture to that of the post-lecture were 23.1% and 21.7% for the General G and the Morality G respectively. Furthermore, there were not students in the General G who could improve their score over 2 points. In contrast, in the Morality G, there were students who could

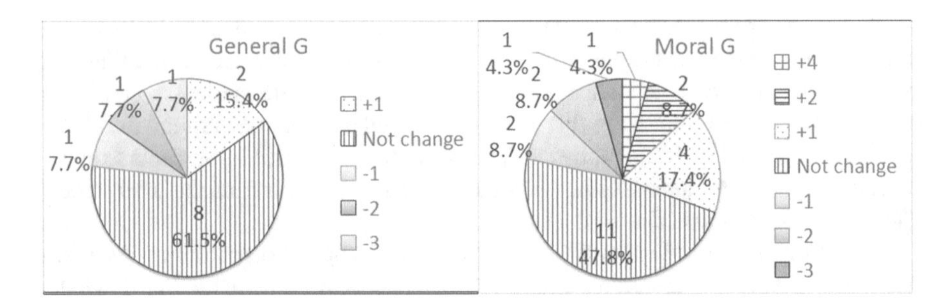

Fig. 2. Correct answer rate of the information ethics examination.

improve their score over 2 points. As a result, the ratio of students whose grade would improve was 15.4% and 30.4% for the General G and Morality G respectively. The morality lecture might be able to increase the number of correct answers to the examination of information ethics. A student in the Morality G who improved her score 4 points could answer correctly to the five questions in the examination of pre-lecture. We could see similar examples. Another student increased her score from 3 to 5, and there was a student who increased her score from 4 to 6.

We analyzed which questions affect the students' score. The students who predominated over the others did not change their score between examinations of pre- and post-lectures. There were two questions to that all of the students could answer correctly. It means that the number of incorrect answers and correct answers is the same.

According to the result of the Morality G, there were two questions to which the students who answered incorrectly before the lecture could correct their answers after the lecture. It must be an effect of our moral education.

Especially, Q7 and Q10 shew significant results. Figure 3 represents the characteristics of the answers' alternation of Q7 and Q10. The contents of Q7 and Q10 were shown in Table 2. For Q7, four students of the Morality G altered their answers from the correct answers to the incorrect answers. But only one student of the General G altered her answers incorrectly. For Q10, five students of the Morality G who answered incorrectly could alter their answers correctly. In contrast, there were only two students

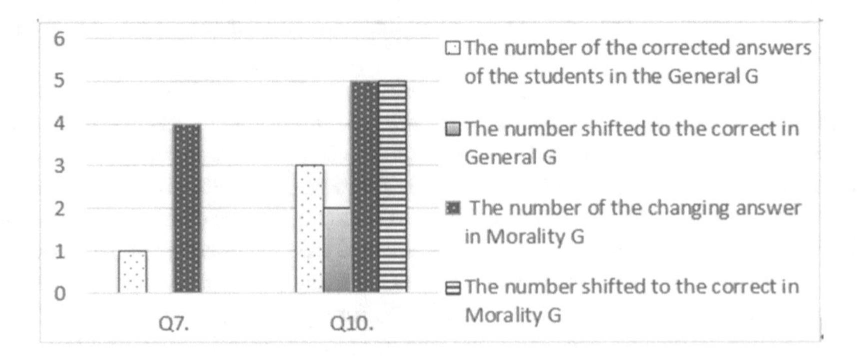

Fig. 3. The characteristics of the students who are the number of correct answers were the same.

who could alter their answers correctly in the General G. The students might be able to correct their answers by the course of time.

5 Discussion

A threat to internal validity is that the developmental stage of students' morality already contains some four or more stage. If it was so, MEA was not successful. It is not necessary to step the self-debate course for students in order to do high-order morality. However, by this research, it was coped with by comparing the change of the number of correct answers of information ethics problems to this threat into Morality G which performed MEA, and General G which performed the general lecture. From the result of Fig. 2, there were four or more stages of moral development of many 2019 students before the moral lecture. Although we did not test 2018 student for moral development stage, it is natural that they have 4 stage (Fig. 3).

However, there is the discrepancy between General G and Morality G in the information ethic problem to the rise of the correct answers. The Self-debate to argue against the reverse opinion of their opinion brought up students in higher morality. Thus, MEA succeeds. A threat to internal validity is that the number of correct answers increases in same problems by vacating time. If it is right, it is not necessary to carry out moral education. However, by this research, it was coped with by investigating the factor of General G who lengthened the score of information ethics problems to the threat. One person was going up from 6 points to 7 points among 10-point full marks, was mistaken in one question compared with a lecture before, and answered two questions correctly. The other was going up from the 7 point to 8 points, was mistaken in two questions compared with a lecture before, and answered three questions correctly. It was shown that the number of correct answers may rise about 15% of the whole problem as time passes. Now, therefore, about 15% of the number of correct answers in Morality G rose may be based on time passes, and it is considered that remaining about 15% is the result by moral education.

However, in Fig. 3, when Morality G received the greeting or some guerdon in the story, all members of the correct answer changed to choose the incorrect answer. If students receive the remuneration of a small sum, and thoughtfulness, they will tend to mistake judgment of information ethics.

6 Conclusion and Future Work

As information ethics education, the method of MEA has been proposed by this research. However, this research analysis is insufficient. For example, in Morality G, the reason for having lengthened the number of correct answers greatly cannot be solved by students with few correct answers giving the lecture (Fig. 3). We want to study deeply information ethics and MEA by the data.

References

1. Ministry of Education, Culture, Sports, Science and Technology.: Fifth Chapter Information Moral Education (2017). http://www.mext.go.jp/b_menu/shingi/chousa/shotou/056/shiryo/attach/1249674.htm. (in Japanese)
2. Ministry of Education, Culture, Sports, Science and Technology.: Fifth chapter Cooperation with Information Moral Education, and the Home and the Community in Schools (2017). http://www.mext.go.jp/b_menu/shingi/chousa/shotou/056/gijigaiyou/attach/1259397.htm. (in Japanese)
3. Okada, Y.: Information ethics education required for college students. Inf. Educ. Center Bull. **23**, 23–26 (2015). (in Japanese)
4. Ministry of Education, Culture, Sports, Science and Technology.: Junior High School Government-Guidelines-for-Teaching description, Special subject Volume on Morality (2017). http://www.mext.go.jp/component/a_menu/education/micro_detail/__icsFiles/afieldfile/2016/01/08/1356257_5.pdf. (in Japanese)
5. Kohlberg, L.: Development of Morality and Moral Education. Hiroike Institute of Education Publishing Dep, Tokyo (1987). (in Japanese)
6. Murakami, F., Inagaki, S.: Analysis of learning materials for information ethical dilemma. In: University ICT Promotion Meeting Annual Conference, National Kyoto International Hall, Hiroshima, pp. 20–27 (2016). (in Japanese)
7. Inagaki, F., Sho, S., Nagato, T., Sumiya, F., Nakamura, F.: The issue of dilemma in the first time information ethics education. In: University ICT Promotion Meeting, Annual Conference Memoirs in 2012, Hiroshima, pp. 43–48 (2012). (in Japanese)
8. Ministry of Education, Culture, Sports, Science and Technology.: Government-Guidelines-for-Education "Power of Living" (2018). http://www.mext.go.jp/a_menu/shotou/new-cs/1387014.htm. (in Japanese)
9. Colby, A., Kohlberg, L.: The Measurement of Moral Judgment, vol. I. Cambridge University Press, Cambridge (1987)
10. Araki, N.: Morality Class of The Heated Discussion to do with The Morals Dilemma Teaching Materials Edited for Junior High School and High School, Morality Development Study Group, Takamatsu (2013). (in Japanese)
11. Araki, N.: Moral Dilemma Materials in Morality Development Study Group, Kokushikan University Physical Education and Sports science society, Tokyo, pp. 27–34 (2010). (in Japanese)
12. Shibazaki, B.: The Sympathy as The Morality Fact - One Preliminary Essay about. The Establishment Opportunity of Act, Sophia University Philosophy Meeting Philosophy Parnassus, vol. 17, pp. 35–49, Tokyo (1988). (in Japanese)
13. Kina, H., Maehara, T.: A Study of "Presentation Behavior" in A Moral Class, Ryukyu University Education Dep. Educational Practice Research Instruction-Center Bulletin, vol. 6, pp. 119–138, Okinawa (1998). (in Japanese)
14. Abe, K.: An Attempt to Establish and Utilize LMS of a Smart-Phone, Higashichikushi Junior College Bulletin, vol. 47, pp. 125–141 (2016). (in Japanese)
15. Lesson support teaching materials "Information ethics PowerPoint drill". Zitsumu Practical Publishing (2017). http://www.jikkyo.co.jp/download/detail/59/322061207. (in Japanese)

Notification Messages Considering Human Centered Design

Junko Shirogane[1(✉)], Yukari Arizono[2], Hajime Iwata[3], and Yoshiaki Fukazawa[2]

[1] Tokyo Woman's Christian University, Tokyo, Japan
junko@lab.twcu.ac.jp
[2] Waseda University, Tokyo, Japan
[3] Kanagawa Institute & Technology, Kanagawa, Japan

Abstract. Messages to notify users of application states (hereafter notification messages) such as error, warning, confirmation, and information messages are often used in user interfaces. Generally, error messages are the most critical, while information messages are the least. However, the types of notification messages are determined in terms of what the application can or cannot process successfully instead of what the users feel are the most critical issues. Currently, human centered design (HCD) focuses on users' aspects. HCD emphasizes users in application development to realize a high usability. Thus, we propose a strategy to realize notification messages considering HCD. Concretely, we analyze existing notification messages in terms of the criticality level by users and define new types of notification messages. In addition, we develop guidelines to implement notification messages effectively. Our strategy allows users to be appropriately notified based on the criticality.

Keywords: Notification message · Human centered design
User interfaces

1 Introduction

Messages in applications (hereafter called notification messages) allow users to understand the application states. These notification messages are classified into types (message types), enabling users to recognize the criticality levels of the application states. Although there are various definitions of notification message types [1–3], they are generally grouped as error, warning, confirmation, and information messages. An error message indicates that the process cannot continue due to a system error or user mistake. A warning message indicates that the process may cause some problems. A confirmation message indicates that the application requires confirmation from the user. An information message notifies the user of something. The most critical message type is an error message, while the least critical is an information message. Thus, notification messages are classified into message types in terms of whether processes can continue successfully or not.

M. Virvou et al. (Eds.): JCKBSE 2018, SIST 108, pp. 53–63, 2019.
https://doi.org/10.1007/978-3-319-97679-2_6

Meanwhile, human centered design (HCD) is widely known to realize highly usable applications [4,5]. Previously, applications were developed in terms of applications' aspects. In these developments, developers focus on application functions and system structures. However, users' real requirements and troubles were not clarified, rending the developed applications unusable. Today, many users without sufficient computer knowledge are operating applications. Consequently, the importance of usability has been increasing. In HCD, applications are developed by focusing on users' aspects. Thus, developers can identify users' real requirements and issues, leading to usable applications.

With respect to HCD, notification messages should notify users of states so that they can recognize the criticality levels in terms of users' aspects [1]. However, notification messages are designed in terms of applications' aspects, not the users' aspects. For example, when a user makes a mistake inputting their password in the login process, an error message is shown. Because the password can be reentered and the login process can continue successfully, a password mistake is not critical to the user. On the other hand, a confirmation message is shown when the user overwrites a file. If the user accidentally overwrites a file, the existing file is deleted. This file deletion is critical to the user. Thus, the current notification message types do not represent the criticality levels that users feel.

To represent notification messages with appropriate criticality levels with regard to users, we propose a new strategy to implement notification messages. In our strategy, we define new message types and develop guidelines to implement notification messages based on graphical user interfaces (GUIs). Our strategy realizes more careful notification messages, allowing users to recognize more critical states. Consequently, fatal states can be avoided.

The rest of this paper is organized as follows. Section 2 describes related works. The newly defined message types are defined in Sect. 3, while Sect. 4 shows our strategy in detail. Section 5 evaluates our method and discusses the results. Finally, Sect. 6 concludes this paper.

2 Related Works

Several kinds of researches have been conducted for notification messages. For improvement of message design, Kelkar et al. aim to prevent users from being confused by inappropriate error messages and developed a tool to determine designs of error messages [6]. They define five categories of error messages, such as irrecoverable errors, faulty use-case errors, absent use-case errors, faulty user input errors, and temporary system errors. Developers select categories of the target error messages and answer questionnaires related to the categories. Then button combinations and icons, title colors are determined by the tool. Explanatory visualization is proposed to represent compile errors visually [7]. Because existing messages of compile errors do not represent enough information that developers can understand the compile errors, they aim to convey messages of compile errors in detail. This method represents messages of compile errors with starting points of the compile errors, points where the compile errors indicate, their associations, and explanatory codes. Although these researches contribute to improve usability of notification messages, message types are not considered.

For usability of notification messages, Inal et al. consider that developers' mindset affect designs of error messages and investigate the factors [8]. They ask subjects (the actual developers) styles of error messages, places to display them, explanations, and the reasons. Developers' preferences are clarified. Sadiq et al. evaluate usability or error messages in Enterprise Resource Planning (ERP) systems and identify usability problems [9]. The evaluations are performed by two methods; evaluations based on 10 heuristics of Nielsen [10] by evaluators and usability tests by users. These researches preferences and problems of error messages, however, concrete design strategies are not considered.

In addition, there are some researches about influences to users from notification messages. Silic investigate How colors of warning messages affect users' decision-making processes [11]. For warning messages with several colors, such as black, blue, yellow, red, green, and white, progression frequencies and time of subjects in United States and India are measured, and the influences are analyzed. Cultural differences are also analyzed. Barik et al. investigate usefulness of messages of compile errors [12]. The points of the investigation consist of effectiveness and efficiency of the messages, whether developers read the messages, and whether the messages make resolution of the compile errors difficult. Subjects resolve the actual compile errors in the experiments, line of sights and video are recorded, and subjects answer questionnaires. These data and code that subjects modify are analyzed. Although these researches are important in terms of how users use notification messages, appropriateness of message types are not mentioned.

3 Human Centered Message Types

To show notification messages with the appropriate criticality, we define four message types based on the users' criticality level. We call these message types "human centered message types" (HCMTs). To define HCMTs, we surveyed notification messages of numerous applications, including office applications, web browsers, and drawing applications. Then notification messages were classified by types based on importance such as loss of something important (data, money, etc.), unachieved purpose, difficult to redo, easy to redo, achieved purpose, and information provision. Meanwhile, we also surveyed typical crucial states by referring to [1,13]. Based on the survey results, we defined four HCMTs. It should be noted that HCMTs represent the criticality levels for the users, but not whether the states already occurred or will occur.

High. States where the user has already lost or is highly likely to lose something such as data or money

Middle. States where the user's operation does not or will likely not finish completely

Low. States where the user can redo the operation easily (e.g., an input mistake) or a problem that does not inhibit the application process
Information. Information provisions of useful information to the user

Table 1 shows some example of notification messages, their HCMT classification, and the justification of the classification.

Table 1. Examples of HCMT notification message classifications

Example notification message	HCMT	Reason
Confirmation to overwrite a file	High	If the user overwrites a file by mistake, important data is deleted
Failure to open a file	Middle	Although the user cannot complete their operation, a critical state such as data deletion does not occur
Input mistake of password	Low	The user can easily reenter their password, and a crucial state does not occur
Provision of security affect information	Information	This is provisional information only, and does not user operations or the application processes

4 Guidelines to Show Notification Messages

Notification messages must be effectively displayed to the users. Often a notification message is shown with a corresponding icon. Although HCMTs represent criticality levels as suggested by the notification message, they do not indicate the necessity and urgency levels of the user's operation. Thus, we developed guidelines to show notification messages. The guidelines include icon designs, styles to show notification messages, commit button usages, sound usages, and text representations corresponding to the target HCMTs.

4.1 Icons of Notification Messages

We designed icons for HCMTs. Often color is used to denote the criticality level. JIS Z 9103 defines the meanings of red, yellow, and green for safety signs and safety display as [14]:

Red. Fire protection, prohibition, stop, danger, and urgency
Yellow. Warning, existence notification, and caution
Green. Safety and progress

In our icons, red means dangerous and urgent, while green means less dangerous and urgent, and yellow is in between. Based on these color meanings, red, yellow, and green respectively indicate high, middle, and low HCMTs.

Additionally, blue denotes information, which we selected after referring to user experience guidelines of Microsoft Windows (Windows guidelines) [15]. Figure 1 shows our icons.

Fig. 1. Icons of HCMTs

4.2 Notification Message Styles

We defined guidelines to represent the necessity and urgency (HCMT style guidelines) by referring to the reference [16], Windows guidelines [15], and Human Interface Guidelines for Apple platforms [17]. The guidelines include five different notification message styles:

Dialog box. An urgent user operation is necessary. A window for the notification message is shown.
In-place. Data must be re-inputted due to an input or selection mistake. The notification message appears at the location where a mistake is identified.
Balloon. Although a user's operation is not required urgently, problems for the process may occur. The notification message is shown in a balloon without a close button.
Notice. Even if the user does not acknowledge the notification message, the application can run safely. The notification message is shown in a balloon with a close button.
Banner. A problem is/was prevented. The notification message is shown at the top of the window.

Table 2 shows the styles. √ indicates that a style can be used for specific types of HCMT.

Table 2. Possible styles to HCMTs

		HCMT			
		High	Middle	Low	Information
Style	Dialog box	√	√	n/a	n/a
	In-place	√	√	√	n/a
	Balloon	√	√	√	√
	Notice	n/a	n/a	√	√
	Banner	√	√	√	√

Figures 2, 3, 4, 5 and 6 show examples of the notification message styles. The notification messages in the dialog, balloon, and notice styles are shown by windows or balloons, and a user can easily recognize the notification message. However, the in-place and banner notification message styles are shown in the windows, but a user may have difficultly recognizing them. Thus, the notification message has a background color that is a lighter shade of the HCMT icon color.

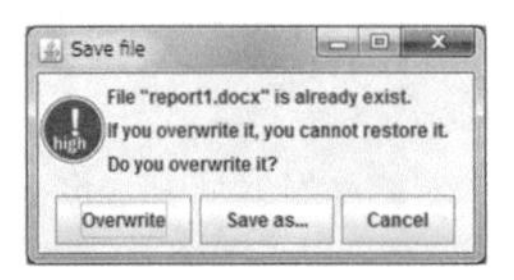

Fig. 2. Example of the dialog box style

Fig. 3. Example of the in-place style

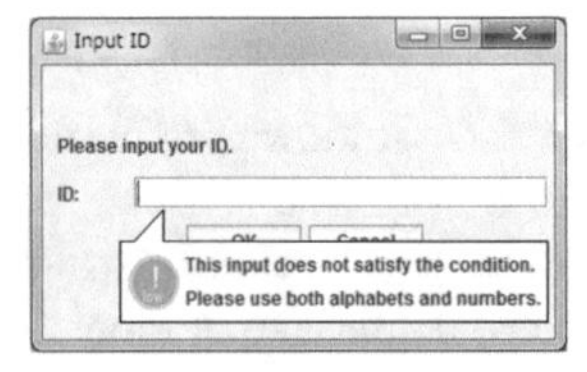

Fig. 4. Example of the balloon style

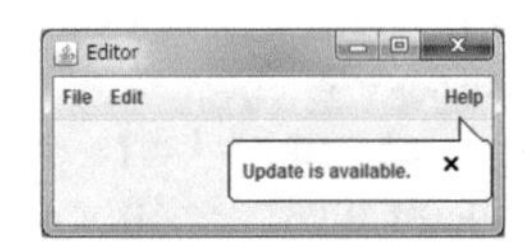

Fig. 5. Example of the notice style

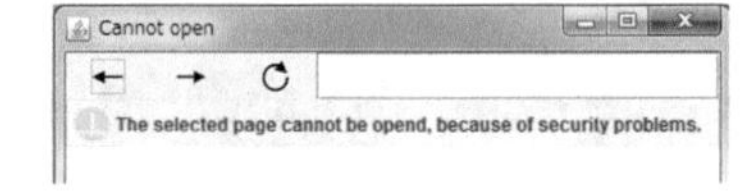

Fig. 6. Example of the banner style

4.3 Text Representation

Notification messages show states that may occur in the future (future states) and those that have already occurred (past states). The text in the notification message indicates the actual state. Notification messages for future states include:

- Problems that may occur
- Operations that can avoid the problems

For the past states, notification messages include:

- Problems that have already occurred
- Operations that can resolve the problems (if available)

4.4 Commit Button Usages

The future and past states are also represented by commit buttons. Commit buttons proceed or stop processes and convey the user's intents to the application. Examples include "OK", "Cancel", and "Close". In the future states, the process should be cancelable. Thus, buttons to proceed with the process and "Cancel" buttons are arranged as commit buttons. In addition, to avoid proceeding with a process by mistake, the "Cancel" button is set as the default button. For past states, only "Close" buttons are arranged because the states represented by the notification messages already occurred.

4.5 Sound Usages

Sounds can effectively notify users of urgency, but they also induce anxiety in the users. If sounds are used in various notification messages, their effectiveness decreases. Thus, sounds are used for notification messages with HCMT of "high", which are especially urgent.

5 Evaluation

To evaluate the appropriateness of our HCMTs, we conducted experiments.

5.1 Experiment Design

We developed two applications. In the first one, notification messages are classified into our HCMTs and shown along with HCMT style guidelines (HCMT style application), while in the other, notification messages are shown using Windows guidelines [15] (existing style application). The only difference between the two applications are the notification messages. The other features such as function and operation are the same.

We defined the operation flow to be operated by subjects. Then the subjects followed the operation flow and then answered questionnaires. The operation flow involved eight steps and included several notification messages. In the steps shown below, parenthesis denote the type of notification messages shown after a subject operated a step. Message types of the notification messages in HCMT and existing style applications are described as "HCMT" and "existing", respectively.

1. Input an invalid password. (HCMT: low; existing: error)
2. Open a file with a file type that is not supported. (HCMT: middle; existing: error)
3. Edit text.
4. Save a file to an existing file. (HCMT: high; existing: warning)
5. Edit text again.
6. Insert an image with a file type that is not supported. (HCMT: middle; existing: error)
7. Exit the application without saving the file. (HCMT: high; existing: warning)
8. Save the file to an existing file. (HCMT: high; existing: warning)

The subjects were 14 university students. Seven subjects operated the HCMT style application followed by the existing style application, while other seven subjects operated the existing style application followed by the HCMT style application. Then the subjects answered questionnaires about the notification messages. For each notification message, the questionnaire asked about the criticality level of the state, the appropriateness of the notification message style, and a free comment. Except for the free comment, the subjects answered the questions on a five-point scale where 1 indicated the least critical or appropriate and 5 indicated the most critical or appropriate.

5.2 Result and Discussion

Figures 7 and 8 show the results for the criticality levels, while Figs. 9 and 10 show the results of the appropriateness of style. The figures show the number of subjects by criticality or appropriate levels for each notification.

Because the subjects answered questions about the same criticality levels in both applications, Figs. 7 and 8 should produce the same results. However, there are differences because the subjects answered questions for every notification message. If subjects answered the questions after operating the applications, they may forget their feelings about the notification messages. Consequently, the results differ. The averages for the criticality levels of notification messages for steps 1, 2, 4, 6, 7, and 8 are 2.46, 4.21, 4.71, 4.11, 4.71, and 4.71, respectively. As described in Sect. 5.1, the criticality levels for these steps in HCMT are "low", "middle", "high", "middle", "high", and "high", respectively. Although the criticality levels of steps 2 and 6 are slightly higher, the criticality levels of the steps by subjects are similar to our classification using HCMTs. According to the free comments, some subjects had difficultly recognizing errors and warnings by the HCMT classifications of the notification messages because they were familiar with existing classifications. To resolve this problem, it is better to add descriptions of existing classifications to the states in the notification messages.

In questions about the appropriateness of the notification message styles, we analyzed the responses for each step by a t-test [18]. Significant differences were not observed at the 95% confidence interval. That is, the subjects did not consider our styles to be more appropriate than existing notification message styles. According to the free comments, some subjects were unfamiliar with green icons and had difficultly reading the texts in the icons due to the lack of color contrast. In addition, some subjects were unable to intuitively understand the icon meaning. To resolve these problems, the colors and shapes of the icons must be improved.

Although the notification message styles have some problems, HCMTs appropriately classify the states, confirming the effectiveness of our HCMTs.

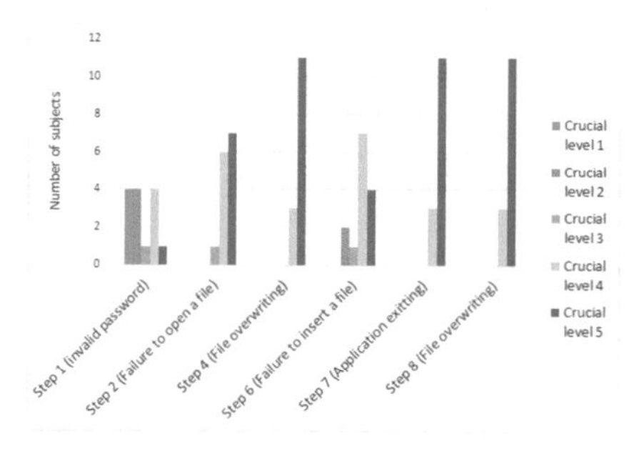

Fig. 7. Criticality levels by subjects' answers for the HTMC style application

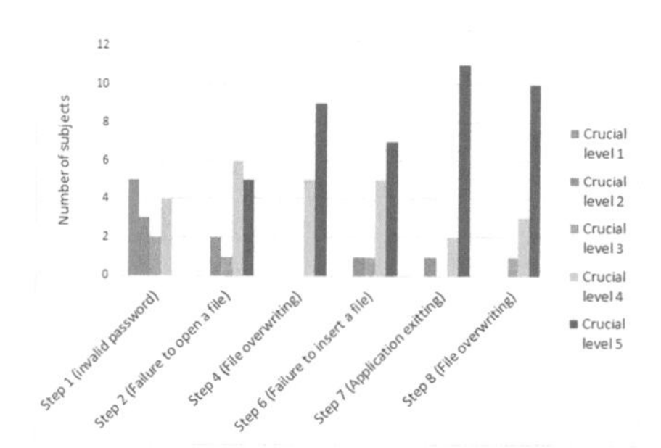

Fig. 8. Criticality levels by subjects' answers for the existing style application

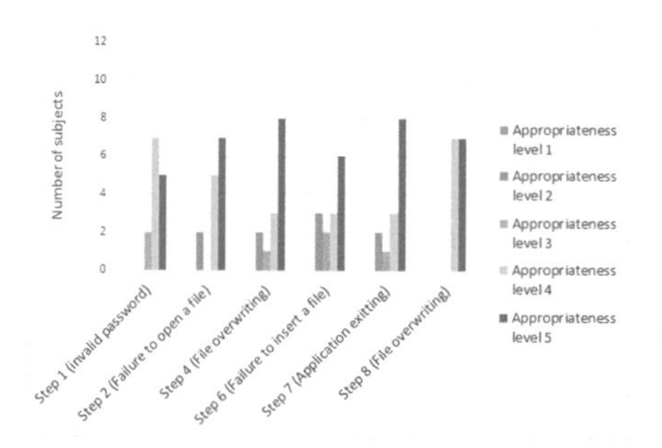

Fig. 9. Appropriate levels by subjects' answers for the HTMC style application

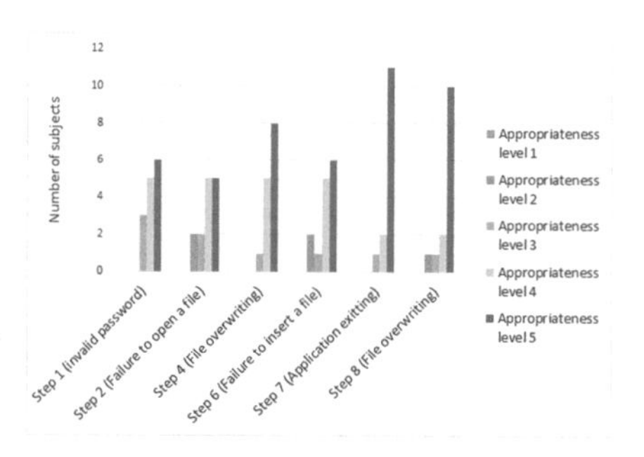

Fig. 10. Appropriate levels by subjects' answers for the existing style application

6 Conclusion

Current message types are determined based on applications' aspects. These message types do not represent the criticality levels that users feel. Thus, to represent notification messages based on users' aspects, we propose new message type called HCMT. HCMTs are divided into four levels: high, middle, low, and information. High represents the most critical states for users, while information represents the least critical.

In addition, we developed guidelines to show notification messages by HCMTs effectively. The guidelines include icon design strategies, notification message styles, commit button usages, sound usages, and text representations. We evaluated the appropriateness of the notification message classifications by HCMTs and the notification message styles. Although some issues are noted with the styles, the notification message classifications using HCMTs are appropriate.

Future work includes:

- Improving icon design, such as color and shape.
- Representing of errors and warnings in notification messages along with HCMTs.
- Generating programs of notification messages along with HCMTs based on specifications.

References

1. Cooper, A., Reimann, R., Cronin, D., Noessel, C.: About Face: The Essentials of Interaction Design. Wiley, Indianapolis (2014)
2. Reiss, E.: Usable Usability: Simple Steps for Making Stuff Better. Wiley, Indianapolis (2012)
3. Constantine, L.L., Lockwood, L.A.D.: Software for Use: A Practical Guide to the Models and Methods of Usage-Centered Design, Addison-Wesley Professional (1999)
4. ISO 9241-210:2010, Ergonomics of human-system interaction - Part 210: Human-centered design for interactive systems (2010)
5. Still, B., Crane, K.: Fundamentals of User-Centered Design: A Practical Approach. CRC Press, Boca Raton (2017)
6. Kelkar, T., Gadepalli, P., Indurkhya, B.: An assistive tool to improve usability of error dialogs. In: Proceedings of 2013 Conference on Technologies and Applications of Artificial Intelligence (TAAI 2013) (2013)
7. Barik, T., Lubick, K., Christie, S., Murphy-Hill, E.: How developers visualize compiler messages: a foundational approach to notification construction. In: Proceedings of 2014 Second IEEE Working Conference on Software Visualization (VISSOFT 2014) (2014)
8. Inal, Y., Ozen-Cinar, N.: Achieving a user friendly error message design: understanding the mindset and preferences of turkish software developers. In: Proceedings of 5th International Conference on Design, User Experience and Usability, Held as Part of the HCI International 2016 (DUXU 2016) (2016)
9. Sadiq, M., Pirhonen, A.: Usability of ERP error messages. Int. J. Comput. Inf. Technol. **03**(05) (2014)
10. Nielsen, J.: 10 usability heuristics for User interface design (1995). https://www.nngroup.com/articles/ten-usability-heuristics/. Accessed 26 Mar 2018
11. Silic, M.: Understanding colour impact on warning messages: evidence from US and India. In: Proceedings of the 2016 CHI Conference Extended Abstracts on Human Factors in Computing Systems (CHI EA 2016) (2016)
12. Barik, T., Smith, J., Lubick, K., Holmesz, E., Fengy, J., Murphy-Hill, E., Parnin, C.: Do developers read compiler error messages? In: Proceedings of 2017 IEEE/ACM 39th International Conference on Software Engineering (ICSE 2017) (2017)
13. Neil, T., Tidwell, J.: Mobile Design Pattern Gallery: UI Patterns for Smartphone Apps. Oreilly & Associates Inc. (2014)
14. JIS Z 9103:2005, Safety colours - General specification (2005)
15. Guidelines. https://msdn.microsoft.com/en-us/library/windows/desktop/dn688964(v=vs.85).aspx. Accessed 2 Mar 2018

16. Johnson, J.: Designing with the Mind in Mind, Simple Guide to Understanding User Interface Design Guidelines, 2nd edn. Morgan Kaufmann (2014)
17. Human Interface Guidelines. https://developer.apple.com/macos/human-interface-guidelines/overview/themes/. Accessed 2 Mar 2018
18. Hartshorn, S.: Hypothesis Testing: A Visual Introduction To Statistical Significance. Independently Published (2017)

Requirements Exploration by Comparing and Combining Models of Different Information Systems

Haruhiko Kaiya[1(✉)], Kazuhiko Adachi[1], and Yoshihide Chubachi[2]

[1] Faculty of Science, Kanagawa University, Hiratsuka 259-1293, Japan
kaiya@kanagawa-u.ac.jp

[2] Advanced Institute of Industrial Technology, Tokyo 140-0011, Japan

Abstract. A people or an organization participates in several different activities simultaneously. Some activities are already supported by information systems, and others are not yet supported but supposed to be. In this paper, we proposed and exemplified a method for exploring the efficiency and synergy of the systems by comparing and combining system elements. We use any modeling languages such as use case diagrams and data flow diagrams for this comparison and combination. For the efficiency, elements in a model are shared or substituted by/for those in another. For the synergy, elements in several models are combined so as to resolve the problems of stakeholder. To validate and discuss our method, we applied it to three sets of activities: room and seat booking, driving and taxi, and rainy report and wiper control. Three different notations are used for each case: use case diagrams, activity diagrams and data flow diagrams.

Keywords: Requirements elicitation · Systems of Systems
Modeling languages

1 Introduction

Each people or organization normally participates in several different activities simultaneously. For example, some faculty member is writing technical papers while he is also preparing teaching materials in class rooms and attending administrational meeting in his faculty. Some activities have been already supported by information systems, but others have not. When we are going to introduce new information systems into latter type of activities, we have to take existing systems into account. However, we do not explicitly take them into account. Requirements engineers simply perform requirements elicitation from stakeholders. They can take the existing systems into account indirectly when the stakeholders states other activities and existing systems related to the activities.

Even when all activities are already supported by some information systems, it is valuable to analyze their relationships. Such analysis sometimes brings the

M. Virvou et al. (Eds.): JCKBSE 2018, SIST 108, pp. 64–74, 2019.
https://doi.org/10.1007/978-3-319-97679-2_7

efficiency and synergy of such systems. In some cases, features of the systems such as functions and data are duplicated among several systems. When we share such features, we can save both logical resources such as source codes or data and physical resources such as hardware devices. Even when some features of a system are not directly related to other features of other systems, the combination of such features sometimes lets us know requirements of which stakeholders were unaware. For example, a function of a system supporting technical writing may contribute to teaching preparation so that its teaching materials are always state of the art.

To summarize, we have the following research questions (RQs).

- RQ1: How to share elements among different systems or activities for efficiency?
- RQ2: How to substitute an element in a system or an activity for another element in another system or another activity?
- RQ3: How to discover unaware requirements by examining different systems and activities?

To answer these questions, we propose a requirements elicitation method. In the method, we first describe models of both existing and developing systems respectively. We then compare elements in a model with others in another model to find model elements which can be shared or substituted. We also examine the combination of such elements to discover requirements of which stakeholders were unaware. To examine such comparison and combination comprehensively, we use any kind of modeling languages. We use three different notations in this paper: use case diagrams, activity diagrams and data flow diagrams.

There already exists a research area called Systems of Systems (SoS) [1]. SoS researches normally focus on the relationships among a huge number of systems, and do not focus on specific comparison and combination of systems' features. Our research focuses on a small number of systems because the number of activities where a people or an organization can participate is normal not so huge, i.e. around seven.

The rest of this paper is organized as follows. In the next section, we explain our requirements elicitation method by using use case diagrams. In Sect. 3, we show other examples using activity diagrams and data flow diagrams respectively. We also discuss examples and the method itself. We then briefly review related works in Sect. 4. Finally, we summarize our current results and show our future issues.

2 Method for Eliciting Requirements for Systems Together

Our requirements elicitation method consists of the following seven steps.

1. Identify stakeholders, hardware devices and physical phenomena. We call them *actors*.

2. Each actor usually participates in several different activities. Identify such activities.
3. Describe a model for each activity. If it already contains information systems, the systems of course should be described in the model. Information systems to be developed may be also described if they are planed.
4. Find the same or similar elements among models of several activities.
5. Share the elements among several activities.
6. Substitute existing elements for those to be developed if they are the same or similar.
7. Identify problems of actors which are not resolved. Discover the solution of the problems by combining elements in all activities.

We explain the steps above respectively by using an example.

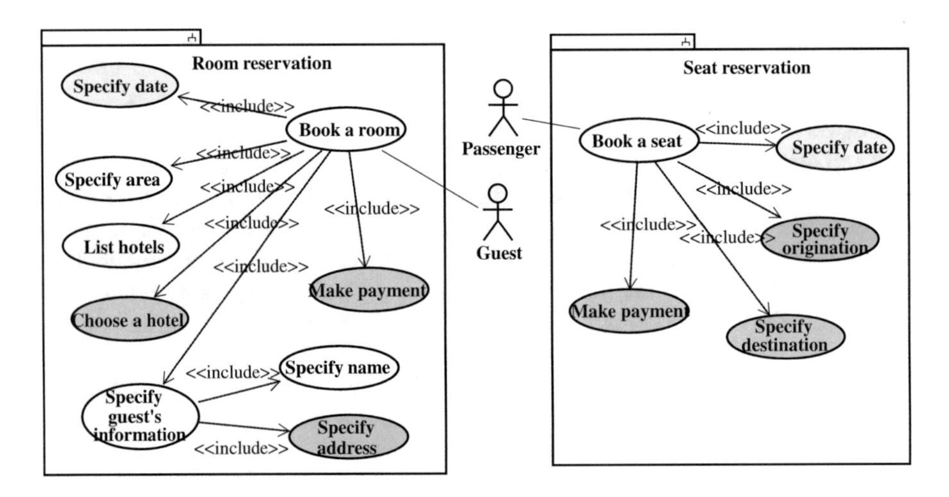

Fig. 1. Use cases of booking a room in a hotel and a seat of a train

2.1 Step 1: Identify Actors

We focus on a person who wants to book a hotel room. A hotel guest is focused here. The left hand side in Fig. 1 shows the use case diagram of this activity.

2.2 Step 2: Identify Other Activities

The person usually has to travel to the hotel by some transportation systems. We thus focus on the activity to book the seat of a transportation system. The right hand side in Fig. 1 shows its use case diagram. For simplicity, we assume the person will board a train for his/her travel. Although we have identified other activities such as checking tourist resorts or booking some activities such as diving or spa, we only focus on two activities here.

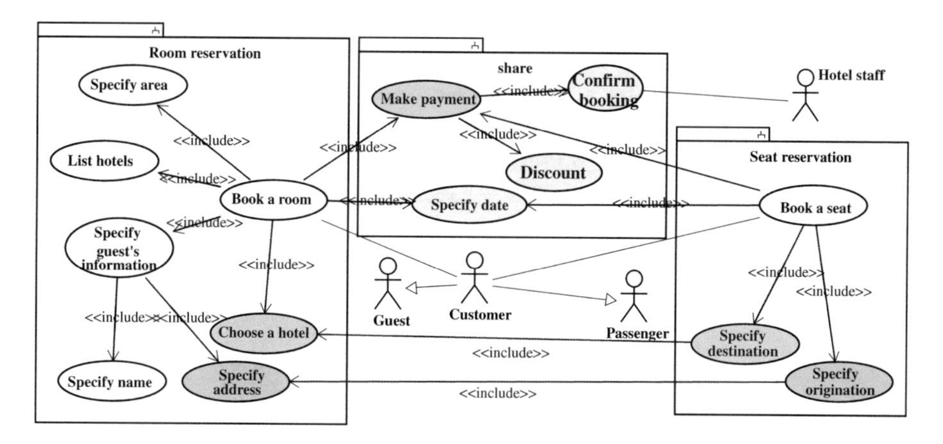

Fig. 2. Sharing elements and discovering new elements of room and seat booking

2.3 Step 3: Describe Models

Use case models are already described in Fig. 1. We may use other models such as class diagrams, state machines or goal models if we cannot find similarity among the models.

2.4 Step 4: Find the Same or Similar Elements

We found two same use cases in these two diagrams: "Specify date", "Make payment". We regard "Choose a hotel" in room booking is similar to "Specify destination" in seat booking because the destination (a train station in this case) is almost fixed according to the chosen hotel. We also regard "Specify address" in room booking is similar to "Specify origination" in seat booking because the origination is usually a station close to the gust's house.

2.5 Step 5: Share the Elements

The same use cases in step 4 are simply shared as shown in Fig. 2. We make include-relationships (or use-relationship) from similar use cases in seat booking to those in room booking.

2.6 Step 6: Substitute Elements

In this case, there is no substitution because no use cases (functions) in a use case diagram cover the functionalities of use cases in another. We mentioned this issue in an example in Sect. 3.2.

2.7 Step 7: Discover New Solutions

Hotel staffs usually participate in this activity even if a booking system (booking web site) exists. No-show is one of the big problems in the booking management of hotel staffs. If both a room and a seat are booked together, hotel staffs can be confident of guest arrival. This new use case "confirm booking" may resolve this problem, and gives peace of hotel staffs' mind. For the customers, the combined system can give some discount because the threat to no-show decreases.

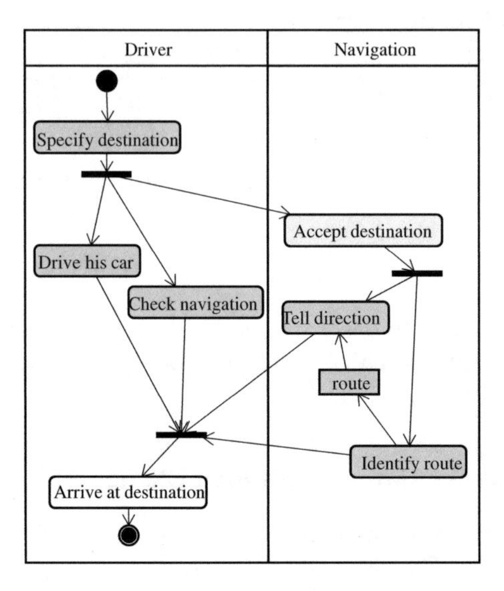

Fig. 3. Activity diagram of a car navigation system

3 Examples

In this section, we show two distinct examples of applying our method to demonstrate the method works. We also discuss our future works on the basis of these examples.

3.1 Driving Navigation and Taxi Call

In this example, we focus on the following two systems, and try to share some actions in the systems.

- Car navigation system for driving (Fig. 3)
- Application for calling a taxi (Fig. 4)

We use activity diagrams because we assume these systems can share several actions so that they can be merged.

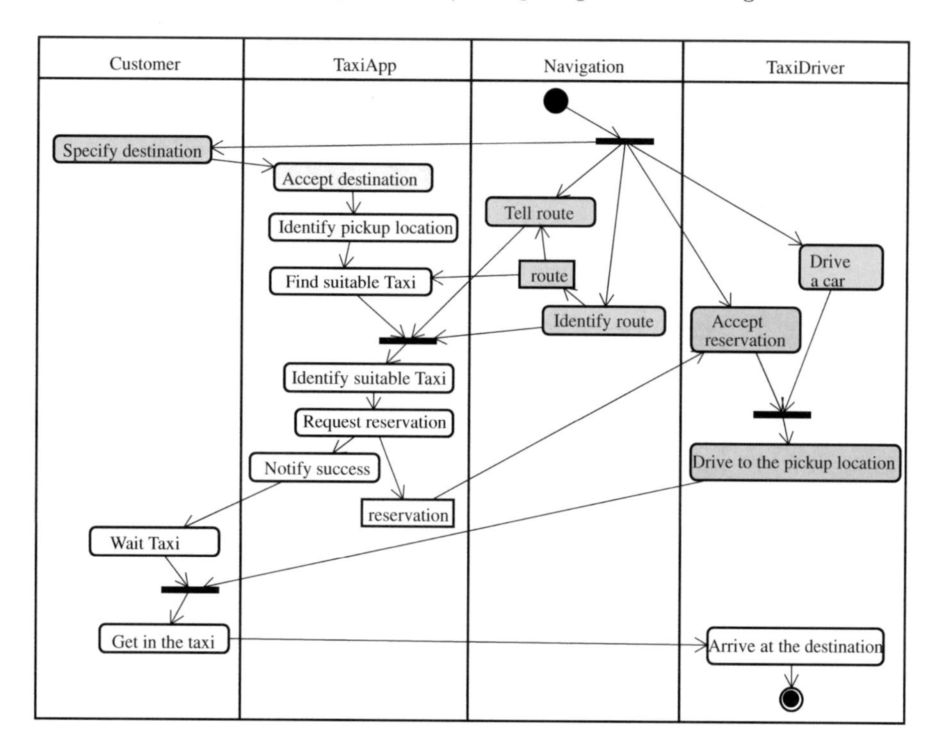

Fig. 4. Activity diagram of an application for calling a taxi

As shown in Fig. 3, a driver first tells his/her destination to his/her navigation system. The system then tells the direction to go. The driver drives his/her car while he/she checks the navigation by the system. The navigation is given by audio and/or visual messages, but such issue is omitted in this diagram for simplicity.

In the context of an application for calling a taxi in Fig. 4, four actors "Customer", "TaxiApp", "Navigation" and "TaxiDriver" exist. We assume the application "TaxiApp" is installed in customers' smartphones, and each customer can call the suitable (normally the nearest) taxi. The application "TaxiApp" communicates with car navigation systems in taxis, and it can identify each taxi's location and vacancy. When the application finds a suitable taxi, it automatically sends a reservation request to the taxi. If the taxi driver "TaxiDriver" may accept the request, the reservation is succeeded. Although the taxi also uses its navigation system after picking its customer up, we omit it in the figure for simplicity.

By comparing these two diagrams, similar elements exists. Actions in "Navigation" are almost the same in the systems. In the navigation system of Fig. 3, the driver performs both actions "specify destination" and "drive a car". In the taxi application of Fig. 4, a customer performs the former action, and a taxi

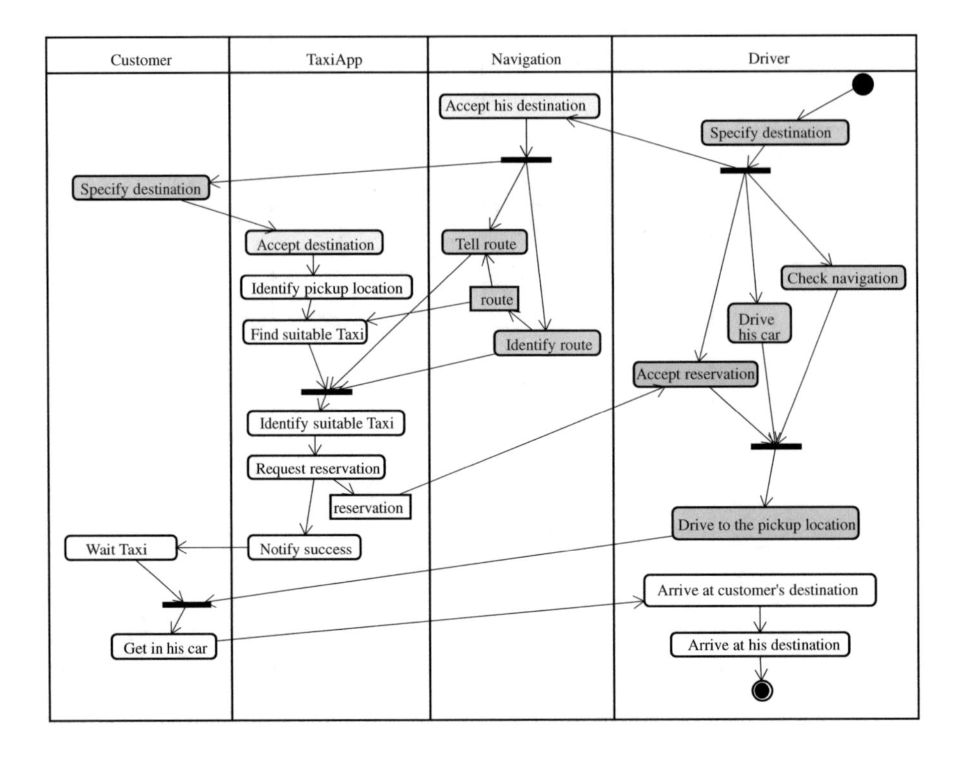

Fig. 5. Activity diagram of an application for asking usual cars as a taxi

driver performs the latter one. We also find the same action "accept destination" in both diagrams, but different actor performs the action respectively.

According to the same actions in "Navigation", we combine these two diagrams into one activity diagram in Fig. 5. In the diagram, we basically substitute "TaxiDriver" for "Driver" in Fig. 3 like Uber[1]. We omit "TaxDriver" in Fig. 5 for avoiding the complexity of the figure, but "TaxiDriver" may continue their works.

As a result, the possibility of catching taxi will increase because usual drivers can also play a role of taxi drivers. Although we have to take security and payment issues into account in the next step, we can find synergy of these two systems by our method.

3.2 Wipers of Cars and Local Weather Report

In this example, we focus on the following two systems. Because we assume the former system for rainy forecast does not exist, we explore how to develop it efficiently.

[1] https://www.uber.com/.

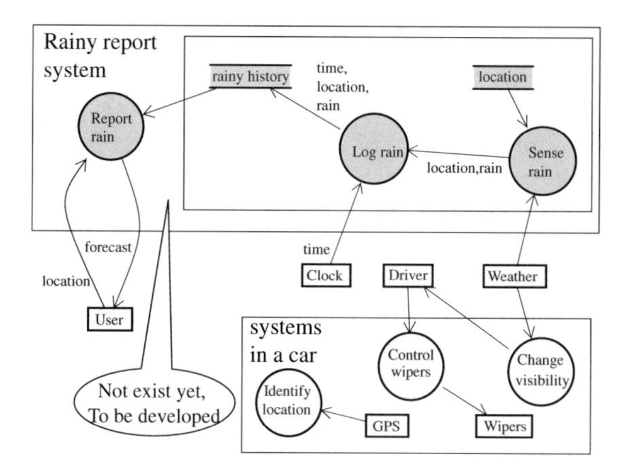

Fig. 6. Rainy report system and systems in a car including wipers control

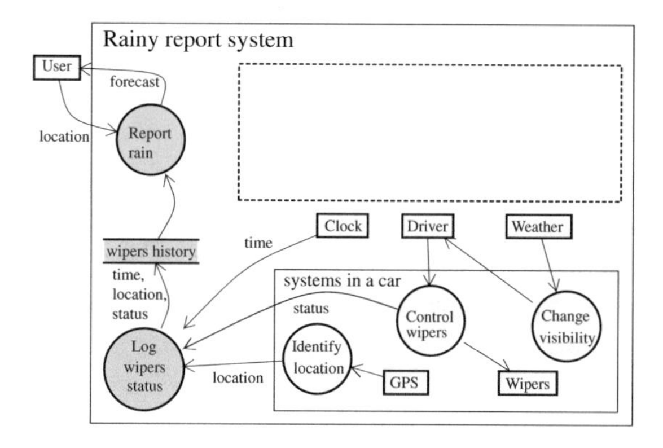

Fig. 7. Substitute wipers status for rain status

- Rainy report system for each small area, e.g. 1 km^2[2], in a big city like London or Tokyo (The system will be developed, i.e. not exist yet)
- Systems in a car such as wipers control and GPS

Recently, heavy rain in a small area becomes a big problem in Tokyo. Many people can keep away from such heavy rain if the former system exists. However, it is hard to forecast such rain in a small area without rainy logs of each small area.

The top area in Fig. 6 shows data flow diagrams of a rainy report system and systems in a car. Although "Sense rain" is represented in a process, it requires some hardware devices for sensing rain. In addition, we have to install many instances of such a process around the city. The location of the process "Sense rain", each rainy status and time are stored in a data store "rainy history". We

[2] The average distance between train stations in a line "Yamanote Line" in Tokyo is about 1.2 km.

assume the cost of developing the system is not low mainly because it requires many devices around the city.

Note that the process "Report rain" forecasts rainy status for each small area on the basis of the rainy history. Each user then gets the forecast of his/her location via his/her smartphone or something like that. We omitted such issues both in Figs. 6 and 7.

Drivers usually uses and controls wipers according to the strength of the rain. The status of the wipers thus reflect the strength of the rain. In addition, GPS system is installed in most cars today for driving navigation. We simply added such systems in a car at the bottom of Fig. 6. The models in the figure let us identify "Weather" activates a process "Change visibility" in the same way as a process "Sense rain". We also identify a process "Control wipers" transitively depends on the "Weather". In addition, car's location can be identified with the help of GPS. Therefore, we can assume "Control wipers" and "Identify location" can record the *quasi*-rainy history. We can thus substitute rain sensors around the city for the cars and their systems as shown in Fig. 7. As a result, we can save the cost for installing sensors around the city. Data flow diagrams in Fig. 6 helps us to find reusable information for the new system.

3.3 Discussion

There are several challenges to perform this method. First, it is not easy to choose the suitable modeling language. Currently, we have to describe models in different notations until we have identified similarities and good combinations. Second, finding the same or similar elements depends on the expertize of requirements analysts. Natural language processing will not be useful because labels and annotations in each model element are normally very short and domain specific. Extending each modeling notation is one of the practical ideas to find them. For example, we may put some semantic tags on each element by using stereo types. Third, focusing on existing problems is not sufficient to discover new requirements. To find needs and desire of which stakeholders are unware, creativity techniques mentioned in the next section can be used but examining such techniques is our future issue.

4 Related Work

There are few approaches for sharing requirements and for exploring unaware requirements by comparing and combining existing or planning systems/services. Instead, we can find methods and techniques for satisfying given requirements by comparing and combining them. In [2], service composition is performed by using goal models so that required functionalities are satisfied. In [3], a requirements analysis method using UML for SoS is proposed. The method also focuses on how to satisfying given efficiency requirements. In [4], a technique for monitoring the deviation from given requirements in SoS is proposed. In a chapter [5], factory automation using SoS is focused. New idea discovery and exploration is not also focused on this chapter. Finding missing requirements is a little bit similar to our research goal. However, most researches [6,7] in the area focuses on the completeness of requirements. In addition, they do not take the synergy of several systems into account.

There already exist creativity techniques [8,9], and some of them have been applied to requirements engineering [10,11]. However, such techniques are not directly applied to models and their elements. Although one research applied one creativity technique to a goal model [12], it only focused on a single system.

Software product line [13] focuses on comparison among systems. However, this research focuses on *similar* systems and their variabilities and commonalities. Our research focuses on comparison and integration among different systems and activities. In a paper [14], integration of different systems and activities were focused. However, the paper only focused on the negative impact of such integration, i.e. security threats.

5 Conclusion

In this paper, we proposed and exemplified a method for sharing, substituting elements among different systems by comparing the elements. We also focused on discover new requirements through the combination of the elements. We use any modeling notations for our comparison and combination. Using different notations helps us to find model elements that can be shared or substituted (Answers of RQ1, 2). For example in a rainy report system in Sect. 3.2, focusing on data flows enables us to find the substitution easily. Problems in an activity are resources of unaware requirements. Because elements in different systems and their combination can become the solution of the problems, our method contributes to the discovery of unaware requirements (Answer of RQ3). However, it seems to be insufficient because not all requirements comes from problem solving. Our future works are summarized in Sect. 3.3 as discussion: choosing the suitable modeling language, finding same/similar elements, discover new requirements beyond the solution of current problems.

Acknowledgement. This work was supported by JSPS KAKENHI Grant Numbers 18K11249, 16H02804, 15H02686, 17K00475 and 16K00196.

References

1. Nielsen, C.B., Larsen, P.G., Fitzgerald, J.S., Woodcock, J., Peleska, J.: Systems of systems engineering: basic concepts, model-based techniques, and research directions. ACM Comput. Surv. **48**(2), 18:1–18:41 (2015)
2. Kritikos, K., Kubicki, S., Dubois, E.: Goal-based business service composition. Serv. Oriented Comput. Appl. **7**(4), 231–257 (2013)
3. Wang, Q.-L., Wang, Z., Liu, Y., Zhu, W.: A novel modeling and analysis approach to efficiency requirements for system of systems. In: International Conference on Oriental Thinking and Fuzzy Logic, pp. 349–360. Springer (2016)

4. Vierhauser, M., Rabiser, R., Cleland-Huang, J.: From requirements monitoring to diagnosis support in system of systems. In: Proceedings of Requirements Engineering: Foundation for Software Quality - 23rd International Working Conference, REFSQ 2017, Essen, Germany, 27 February–2 March 2017, pp. 181–187 (2017)
5. Nahavandi, S., Creighton, D., Le, V.T., Johnstone, M., Zhang, J.: Future integrated factories: a system of systems engineering perspective, pp. 147–161. Springer (2015)
6. Wei, B., Delugach, H.S.: Transforming UML models to and from conceptual graphs to identify missing requirements. In: Haemmerlé, O., Stapleton, G., Zucker, C.F. (eds.) Graph-Based Representation and Reasoning, pp. 72–79. Springer, Cham (2016)
7. Zhang, Z., Thanisch, P., Nummenmaa, J., Ma, J.: Detecting missing requirements in conceptual models. In: Dregvaite, G., Damasevicius, R. (eds.) Information and Software Technologies, pp. 248–259. Springer (2014)
8. Berntsson-Svensson, R., Taghavianfar, M.: Selecting creativity techniques for creative requirements: an evaluation of four techniques using creativity workshops. In: 23rd IEEE International Requirements Engineering Conference, RE 2015, Ottawa, ON, Canada, 24–28 August 2015, pp. 66–75 (2015)
9. Ohiwa, H.: KJ editor for creative work support and collaboration. In: Conference on Creating, Connecting and Collaborating through Computing (C^5), pp. 104–109 (2003)
10. Bhowmik, T., Niu, N., Savolainen, J., Mahmoud, A.: Leveraging topic modeling and part-of-speech tagging to support combinational creativity in requirements engineering. Requirements Eng. **20**(3), 253–280 (2015)
11. Jing, D., Zhang, C., Yang, H.: Using an ideas creation system to assist and inspire creativity in requirements engineering. In: Requirements Engineering in the Big Data Era, pp. 155–169. Springer, Heidelberg (2015)
12. Kinoshita, T., Hayashi, S., Saeki, M.: Goal-oriented requirements analysis meets a creativity technique. In: Advances in Conceptual Modeling, pp. 101–110. Springer (2017)
13. Pohl, K.: Software Product Line Engineering: Foundations Principles and Techniques. Springer, Heidelberg (2011)
14. Kaiya, H., Okubo, T., Kanaya, N., Suzuki, Y., Ogata, S., Kaijiri, K., Yoshioka, N.: Goal-oriented security requirements analysis for a system used in several different activities. In: Advanced Information Systems Engineering Workshops. LNBIP, pp. 478–489 (2013)

Improved Searchability of Bug Reports Using Content-Based Labeling with Machine Learning of Sentences

Yuki Noyori[1(✉)], Hironori Washizaki[1], Yoshiaki Fukazawa[1], Hideyuki Kanuka[2], Keishi Ooshima[2], and Ryosuke Tsuchiya[2]

[1] Waseda University, Tokyo, Japan
akskw-luck@akane.waseda.jp,
{washizaki,fukazawa}@waseda.jp
[2] Hitachi, Ltd. Research & Development Group, Tokyo, Japan
{hideyuki.kanuka.dv,keishi.oshima.rj,
ryosuke.tsuchiya.rs}@hitachi.com

Abstract. Most stakeholders refer to past bug reports when they encounter a problem since bug reports contain useful information. However, searching for specific content is difficult because there are many bug reports. The desired content depends on the viewpoint of the stakeholder. A full text search includes unwanted content, which is costly. Although this problem has been previously noted, a solution has yet to be proposed. Herein we propose Content-based Labeling Method as a solution. This method organizes information in a bug report by labeling each sentence based on its contents, allowing stakeholders' viewpoints to be considered. We evaluate the improvement in searchability. The Content-based Labeling Method improves the searchability according to the F-measure and precision of the experimental results.

Keywords: Bug report · Machine learning · Labeling · Searchability

1 Introduction

Bug reports contain useful information [1] as they describe various kinds of contents, including discussions, phenomena, and solutions. Figure 1 shows an example of a bug report from the Eclipse Bug Repository, which reports defects and their subsequent modifications. Stakeholders search bug reports to resolve encountered problem. However, each stakeholder is interested in different content. That is, each stakeholder wants to search the content for a specific issue. For example, when users encounter a problem, they want to search how to resolve the problem, and they only need to search for "phenomenon". Similarly, when developers want to know the defects for a function, they only need to search for "Cause". Although users and developers search for different content, both have to conduct a full-text search. The number of reports that hit in a full-text search increases. This situation is called low searchability. In our experiment, when searching for one bug report, 10 unrelated bug reports were hit in a full-text search. This problem has been pointed out previously in a study that employed

M. Virvou et al. (Eds.): JCKBSE 2018, SIST 108, pp. 75–85, 2019.
https://doi.org/10.1007/978-3-319-97679-2_8

questionnaires from many reporters and developers [2]. Some people indicated poor searching capabilities, but a practical solution to this issue has yet to be realized. This research aims to narrow the search scope by changing the search according to the viewpoint. We use a labeling method to solve this problem. Because the labeling method considers the stakeholder's viewpoint, our labeling method, called the Content-based Labeling Method, labels each sentence in a bug report as a means to organize the contents. It is necessary to label the sentences according to their contents. If we know the stakeholder's viewpoint or the desired information of the stakeholder, the search results can be tailored.

```
lucas bigeardel2004-07-06 06:31:08 EDT

Group title is not taken into account by Window-Eyes 4.5.
The group title should be pronouced before radio buttons or other controls.
It is not, this is bug since blind users cannot know what the group content is
dealing about.
```
```
Carolyn MacLeod 2004-08-23 17:50:11 EDT

Interesting. Both Window-Eyes and JAWS will read a Group title IF the Group
has a single child that does not have an obvious name, for example, a text or
a list or a combo box.
But neither screen reader will read a Group title if it groups a set of radio
buttons or check boxes, say. (Or, I assume, if it is surrounding something
that has an obvious label to read instead, like a button, and if the child
control takes focus).
We have several dialogs - notably, the search and find/replace dialogs, that
have groups with radio/checkbox children. The screen readers do not read the
group titles in these dialogs.
I will bring this to the attention of the screen reader developers.
For now, an extreme, but sort-of useable work-around in Window-Eyes is to tell
it to read every control in a newly-opened dialog. You can do this with:
Global -> Verbosity -> Activated... Entire Window if Dialog
```

Fig. 1. Example of a bug report; the beginning part of bug 449596 from the Eclipse Bug Repository.

Here we evaluate the searchability and accuracy of the Content-based Labeling Method. The label method adds document structure information peculiar to bug reports as learning data to increase the accuracy of the classifier. In addition, the accuracy of four supervised learning algorithms is compared: namely Support Vector Machine (SVM), Naïve Bayes (NB), Neural Network (NN), and Random Forest (RF). In this research, a multilayer perceptron is used as a neural network, and experiments assess whether the searchability improves. In particular, we investigate two Research Questions (RQs) to evaluate this labeling method:

- **RQ1:** How accurate is our labeling method, which is based on the document structure?
- **RQ2:** How much does the Content-based Learning Method improve searchability?

The rest of this paper is structured as follows. Section 2 discusses related works. Section 3 describes our method, while Sect. 4 presents our experiments. Section 5 considers future work, and Sect. 6 provides the conclusion.

2 Related Works

Several studies have investigated labeling bug reports. Some manually labeled the importance of sentences and implemented a supervised learning approach based on the data [3, 4]. Adding a corpus of bug reports to a corpus applied to a general field improve accuracy of the classifier, and it was found that adding the characteristic feature of the bug report to the teaching data is better.

One study researched labeling using the PageRank algorithm, which is an unsupervised learning algorithm. PageRank was originally an algorithm use to determine the importance of a web page [5]. Another work created a summary through unsupervised learning. Before learning, sentences unrelated to the bugs were removed and a keyword dictionary to weight domain-specific words was created using TF-IDF technique as preprocessing [6]. However, these methods merely judge the importance of each sentence. Unlike these methods, which provide a binary classification of importance, our method classifies multilevel content.

Poor searching capabilities have already been regarded as a problem [2]. According to the results of a questionnaire answered by 22 reporters and 34 developers, the problem is difficulty searching a bug report. Here we present a solution to this problem.

3 Method

Our Content-based Labeling Method is used to consider the viewpoint. Table 1 lists the different stakeholders' viewpoints. No. 1 is the example of a Service User. Typically, a Service User wants to know whether he or she has encountered a bug or not. If a bug is encountered, the Service User has to verify if the bug has been resolved or needs to be fixed. No. 2 and No. 3 are Developer examples. No. 2 is from the viewpoint of relevance. If bugs are related, then they influence each other, and caution must be used when correcting such bugs. No. 3 is from the viewpoint that he or she does not want to search unrelated information. Figure 2 overviews our Content-based Labeling Method, which involves two steps: manual labeling and machine learning. Then we create a classifier to label an unknown bug report. Table 2 shows example bug report sentences corresponding to each label, where "Label" in the left column denotes manual labeling.

Table 1. Stakeholders' viewpoints and labels

No.	Stakeholder	Viewpoint	Label
1	Service User	Is this bug fixed or not?	P
2	Developer	Can related bugs be checked?	P, C, S
3	Developer	Is it possible to search only information related to bugs?	P, C, S, D

P: Phenomenon, C: Cause, S: Solution, D: Discussion

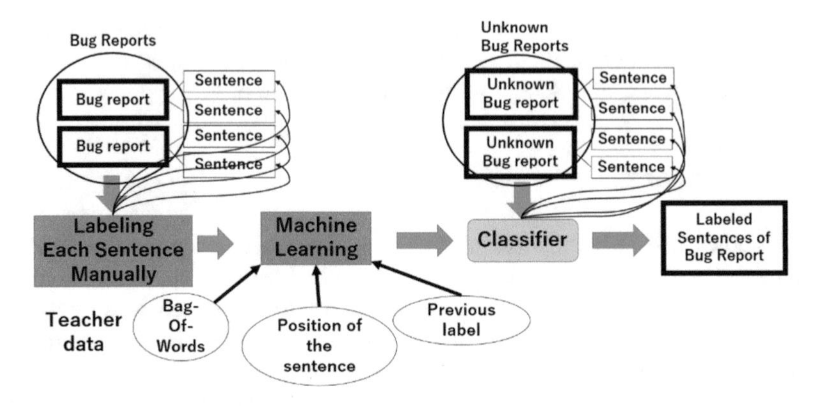

Fig. 2. Overview of our method.

Table 2. Example of a sentence corresponding to each label in a bug report

Sentence	Label
When I try to export reports that contain graphs gnuCash just creates an empty html file	Phenomenon (P)
The problem here is that nsSearchSuffestions.js is passing the wrong previousResult to form history	Cause (C)
Perhaps the page should take CSS from intl.css, where localisers can override some of the classes or add a class to set direction	Solution (S)
Would setting the value to empty-string (|"|") work or something similar?	Discussion (D)
Sounds fine to me. I guess…	Other (O)

Labeling Manually. The contents written in bug reports have been investigated [7]. We prepared our labels based on contents by referencing this study. Too many labels negatively impact the accuracy, while too few labels only allow rough classification. Here we analyzed actual bug reports and discussed the results with industry experts. We prepared five label types:

- Phenomenon (P): P represents the behavior of the encountered bug. Most phenomena are written at the beginning of a report or after someone else has confirmed the bug.
- Cause (C): C is the sentence describing why the bug occurred. In some cases, the cause is specified at the beginning, while in other it is a result of an investigation. Compared to the other labels, this one is less common in the bug report used in the analysis.
- Discussion (D): D is a conversation on how to fix a bug. D includes questions and answers.
- Solution (S): S devises a method to fix a bug such as a patch. Although examples appear in the first half of a bug report, most are in the latter half.
- Other (O): O denotes remarks unrelated to bugs such as greetings and gratitude.

Table 2 shows that viewpoint No. 1 only involves P, while viewpoint No. 2 employs P, C, and S to understand the relationships between bugs. Viewpoint No. 3 uses all labels except O as extraneous information should be excluded. Thus, the necessary labels depend on the viewpoint.

Machine Learning. We created a classifier based on the supervised learning approach. To increase the accuracy, we used the unique document structure and features of bug reports. Previous studies have confirmed that using the corpus of a bug report improves the accuracy of the bug report [3, 4]. In this paper, we use the three features as explanatory variables of teaching data to improve the labeling accuracy:

(1) Bag-of-Words: This is a vector usually used in natural language processing. This vector is expressed by the number of occurrences of words. All words in the document are listed, and the appearance frequency of each word in the list is an element of the vector. This vector ignores the document structure.
(2) Position of sentences: This feature is adopted to learn the characteristics of a bug report. For example, a bug report tends to describe phenomena in the first half and solutions in the latter half. The report in Fig. 1 also starts with the report of phenomena.
(3) The label of the previous sentence: This is used as an explanatory variable to learn the features of a bug report. For example, a report may describe a phenomenon, followed by a discussion and then a conclusion.

To learn this feature, we use four supervised approaches: SVM [8], NB [9], RF [10], and NN [11]. We evaluated and compared the accuracy of these four supervised approaches to select the supervised approach to use in this method. The experiment is conducted in Python. We used the "Scikit-learn" library for machine learning [12] and the "gensim" library to create the Bag-of-Words [13].

4 Evaluation

In this section, we evaluate the accuracy of Content-based Labeling Method and the effectiveness using the two research questions.

4.1 Dataset

We used bug reports freely available [3]. This corpus was created manually by the authors and consisted of bug reports for four Open Source Software (OSS): Eclipse Platform, Gnome, Mozilla, and KDE. There were a total of 36 bug reports and 2361 sentences. IDs are assigned to bug reports 1 to 36 in order. In this research, we assigned labels to 17 bug reports and 982 sentences corresponding to ID.1 to ID.17 in this corpus. Table 3 lists the manual labeling results. There are a few sentences corresponding to C and D. If the number of data points for each label is biased, then the prediction result of the learning model is biased. This is known as a class imbalance problem [14]. We conducted undersampling and oversampling. The Synthetic Minority Oversampling TEchnique (SMOTE) algorithm was used for oversampling [17]. In this study, we randomly selected 186 learning data points as undersampling so that each label employs the same number of data points because 186 is equivalent to the middle value of each label. If a label had less than 186 data points, then all the available data was used. In oversampling, SMOTE was used to match the number of data of each category to the maximum value of 426. We evaluated and compared each undersampling dataset (Without SMOTE) and oversampling dataset (With SMOTE).

Table 3. Results of manual labeling

Phenomenon	Cause	Solution	Discussion	Other
186	27	426	80	263

4.2 Evaluation Metrics

We used accuracy, precision, recall, and F-measure as the evaluation metrics. These are commonly used as evaluation indices for classification problems and searches etc. [15].

(1) Accuracy: Accuracy is the percentage of correctly predicted data for the entire data. This is used to evaluate the machine learning performance.
(2) Precision: Precision is the percentage of actually positive results among the predicted positive results.
(3) Recall: Recall is the percentage of predicted positive results among the actual positive results.
(4) F-Measure: F-Measure is the harmonic mean of precision and recall.

4.3 RQ1: How Accurate Is Our Labeling Method, Which Is Based on the Document Structure?

To evaluate the Content-based Labeling Method, a cross-validation was performed using four machine learning algorithms. We divided each dataset into 10 pieces in this study. The accuracy of each algorithm was compared. This experiment used teaching data, which included the document structure. Figure 3 shows the results of the cross-validation. RF with SMOTE is the most accurate algorithm in this study.

We also compared the cross-validation results when learning used the Bag-of-Words only to that using both the Bag-of-Words and the document structure to verify that the document structure is really an explanatory variable. Figure 4 shows the results. Bag-of-Words only considers the sentence. However, adding the document structure to the training data improves the accuracy for all algorithms without SMOTE because the context of the sentence is considered. With SMOTE, it does not change much, except for SVM. RF has the highest accuracy. The accuracy without SMOTE is about 0.6372, whereas that with SMOTE is about 0.9003. These observations answer RQ1. The RF algorithm is used for the evaluation of searchability. Table 4 shows the labeling results using the RF algorithm for the 19 bug reports that were not used for machine learning. The cause and discussion only have a few descriptions in both of the datasets. Since the number of data is biased and small, the results may be biased.

However, the results show that many bug reports are not well balanced. Instead, general bug reports tend to fix bugs without sufficient discussion. When the example of Fig. 1 is applied to an RF classifier, most sentences of first speaker were judged as a phenomenon, the sentence of the next speaker was judged as solution except for "Interesting". "Interesting" was judged as other.

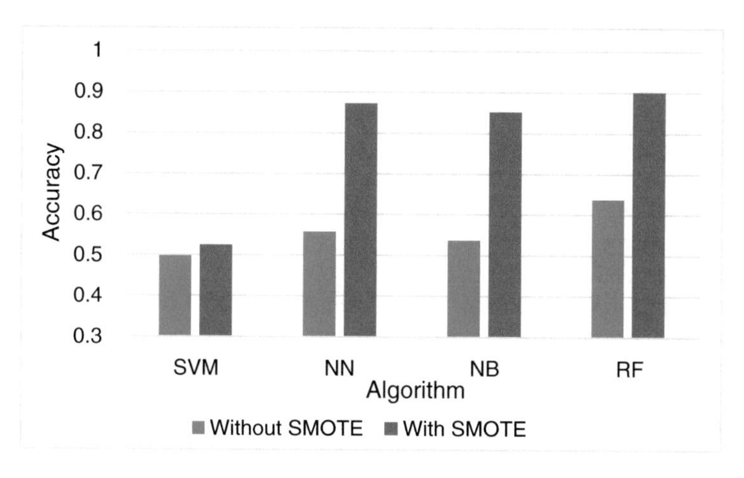

Fig. 3. Cross-validation of the accuracy.

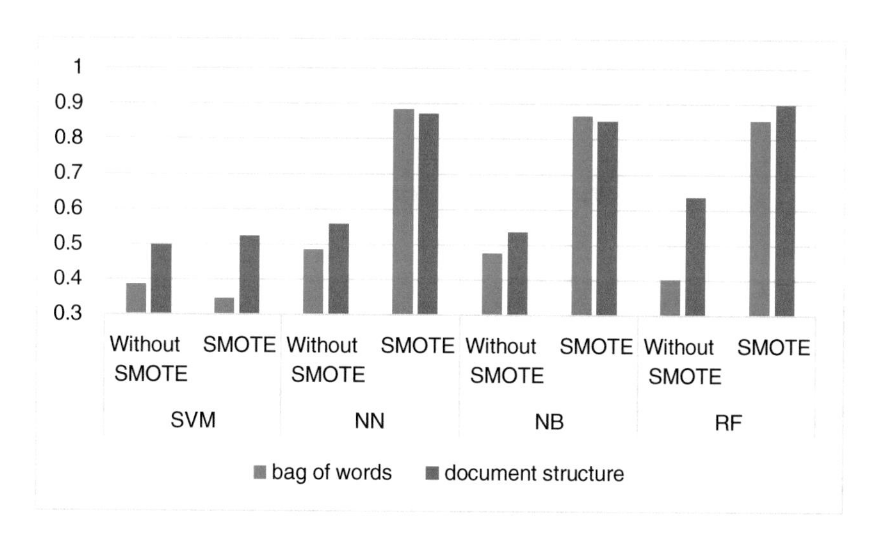

Fig. 4. Comparison of teaching data

Table 4. Results of labeling automatically

	Phenomenon	Cause	Solution	Discussion	Other
Without SMOTE	319	2	65	5	831
With SMOTE	226	2	535	6	453

4.4 RQ2: How Much Does the Content-Based Learning Method Improve Searchability?

We evaluated the increase in searchability to determine whether our approach with the RF algorithm helps improve efficiency. Next, the searchability was evaluated for each viewpoint. We used 19 unique bug reports with ID18 to 36 in this corpus. A keyword such as "keyboard" or "browser" was created based on the title of the bug report and used for searching. We searched for keywords in the sentence of the bug report and compared the number of hits by searching the whole sentences and by searching only sentences corresponding to a specific label.

Table 5 compares part of the results when our method is used and not used to search for phenomena. "Not Applied", "Without SMOTE", and "With SMOTE" are the values without our method, with our method without applying SMOTE, and with our method with applying SMOTE, respectively. ID.22 is an example where applying the method does not change the number of hits. Because only one hit is identified prior to applying this method, improvement is not possible. ID.23 and ID.24 are examples where applying this method improves the number of hits, enhancing searchability. ID.25 is an example where applying this method degrades the number of hits, as even the necessary information is excluded. In Without SMOTE, this may be caused by the poor labeling accuracy due to the small number of bug reports when creating the classifiers. In With SMOTE, over learning occurs because the features of the same characteristic are exaggerated by increasing the amount of data. Hence, necessary information is difficult to acquire if the label differs.

Table 5. Part of the results comparing the search hits

Bug report ID	Search keyword	Method not applied to search hits	Method applied to search hits
ID.22	Calendar	1	1
ID.23	Folder	3	2
ID.24	Select	11	5
ID.25	Keyboard	2	0

We compared the precision and F-measure based on all the keyword hit results according to the viewpoints listed Table 1. The Content-based Learning Method improves the precision and the F-measure for all viewpoints. Without SMOTE has the highest F-measure (Fig. 5). When SMOTE is used, the precision improves but the F-measure decreases. This is attributed to over learning. Increasing the amount of training data should improve the F-measure when SMOTE is used. Without SMOTE, the F-measure improves by about 0.1684. These results confirm that our approach improves the searchability, answering RQ2. For current datasets, our approach without SMOTE is more appropriate it. However, if the amount of training data is large, our approach using SMOTE should be more suitable.

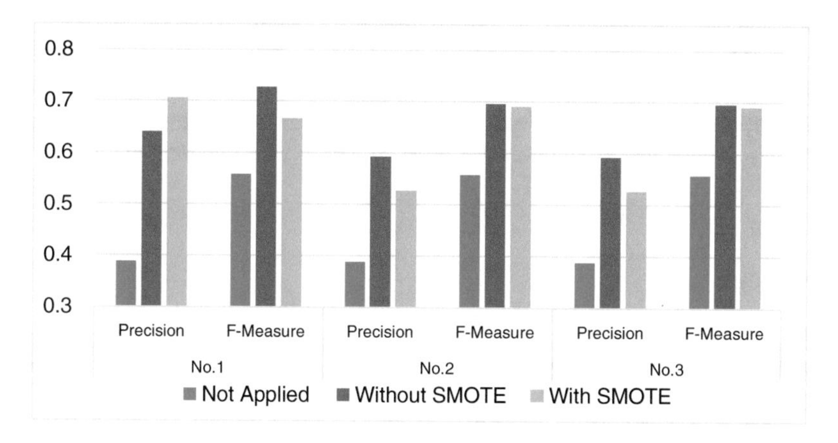

Fig. 5. Comparison of the results

4.5 Threats to Validity

Internal Validity. Threats to internal validity include that we manually created the labels used in the evaluation and that we derived the keywords in the experiments ourselves. Although this study was based on survey results and discussions with industry experts, creating the inputs with multiple experts is a better approach.

External Validity. The number of bug reports was very small. We used bug reports for four projects. Applying this approach to other projects may reduce the accuracy. We should confirm the accuracy and searchability in other conditions.

5 Future Work

To increase the labeling accuracy, we plan to increase the amount of data and add new explanatory variables. Additionally, each stakeholder often has many viewpoints. Currently, we prepared the viewpoints manually. In the future, we want to extract the viewpoint from logs using a topic model similar to Data-Driven Development of Personas [16]. Now, this approach is limited to OSS data. Thus, we are planning to apply this method to industry data. We intend to expand this approach to documents created during software development.

6 Conclusion

In this paper, we propose the Content-based Labeling Method to improve the searchability. The experiment demonstrates that adding the document structure to the explanatory variables increases the accuracy. The RF algorithm with SMOTE provides the best accuracy. We measured the improvement in searchability. For all viewpoints, the Content-based Labeling Method improves both the F-measure and precision. Our approach with SMOTE seems to over learn when the dataset is small. To make this labeling method more practical, our next step is to increase the amount of data and to systematize viewpoint extraction. We want to apply this labeling method to other purposes.

References

1. Bettenburg, N., Just, S., Schroter, A., Weiss, C., Premraj, R., Zimmermann,T.: What makes a good bug report? In: Proceedings of the 16th ACM SIGSOFT International Symposium on Foundations of Software Engineering, pp. 308–318 (2008)
2. Yusop, N.S.M.Y., Grundy, J., Vasa, R.: Reporting usability defects: do reporters report what software developers need? In: Proceedings of the 24th Australasian Software Engineering Conference, pp. 38–45 (2015)
3. Rastkar, S., Murphy, G.C., Murray, G.: Automatic summarization of bug reports. IEEE Trans. Softw. Eng. **40**(4), 366–380 (2014)
4. Rastkar, S., Murphy, G.C., Murray, G.: Summarizing software artifacts: a case study of bug reports. In: Proceedings of the 32nd International Conference on Software Engineering, pp. 505–514 (2010)
5. Ferreira, E.C., Vieira, V., Mourao, F.: Bug report summarization: an evaluation of ranking techniques. In: X Brazilian Symposium on Components, Architectures and Reuse Software, pp. 101–110 (2016)
6. Mani, S., Catherine, R., Sinha, V.S., Dubey, A.: AUSUM: approach for unsupervised bug report summarization. In: Proceedings of the 20th ACM SIGSOFT International Symposium on the Foundations of Software Engineering, pp. 1–11 (2012)
7. Yusop, N.S.M.Y., Grundy, J., Vasa, R.: Reporting usability defects: do reporters report what software developers need? In: Proceedings of the 20th International Conference on Evaluation and Assessment in Software Engineering, pp. 1–10 (2016)
8. Joachims, T.: Text categorization with support vector machines: learning with many relevant features. In: Proceedings of the 10th European Conference on Machine Learning, pp. 137–142 (1998)
9. Zhang, H., Li, D.: Naïve Bayes text classifier. In: Proceedings of the IEEE International Conference on Granular Computing, pp. 708–711 (2007)
10. Wu, Q., Ye, Y., Zhang, H., Ng, M.K., Ho, S.-S.: ForesTexter: an efficient random forest algorithm for imbalanced text categorization. Knowl. Based Syst. **67**, 105–116 (2014)
11. Sebastiani, F.: Machine learning in automated text categorization. ACM Comput. Surv. **34** (1), 1–47 (2002)
12. Scikit-learn machine learning in Python. http://scikit-learn.org/
13. Gensim topic modelling for humans. https://radimrehurek.com/gensim/
14. Garca, S., Herrera, F.: Evolutionary under-sampling for classification with imbalanced data sets: proposals and taxonomy. Evol. Comput. **17**(3), 275–306 (2009)

15. Hripcsak, G., Rothschild, A.S.: Agreement, the F-Measure, and reliability in information retrieval. J. Am. Inform. Assoc. **12**(3), 296–298 (2005)
16. Watanabe, Y., et al.: ID3P: iterative data-driven development of persona based on quantitative evaluation and revision. In: Proceedings of the 10th International Workshop on Cooperative and Human Aspects of Software Engineering, pp. 49–55 (2017)
17. Chawla, N.V., Bowyer, K.W., Hall, L.O., Kegelmeyer, W.P.: SMOTE: synthetic minority over-sampling technique. J. Artif. Intell. Res. **16**, 321–357 (2002)

A Fast Learning Recommender Estimating Preferred Ranges of Features

Takuya Watanabe[1(✉)], Yuji Nakazato[1], Hiroaki Muroi[1], Takuya Hashimoto[2], Toru Shimogaki[3], Takeshi Nakano[4], and Tsutomu Kumazawa[5]

[1] Edirium K.K., 1-8-21 Ginza, Chuo, Tokyo, Japan
{sodium,nakazato,muroi}@edirium.co.jp

[2] Acroquest Technology Co., Ltd., 3-17-2 Shinyokohama, Kohoku, Yokohama, Kanagawa, Japan
hashimoto@acroquest.co.jp

[3] NTT DATA Corporation, 3-3-9 Toyosu, Koto, Tokyo, Japan
shimogakit@nttdata.co.jp

[4] Recruit Technologies Co., Ltd., 1-9-2 Marunouchi, Chiyoda, Tokyo, Japan
tf0054@r.recruit.co.jp

[5] Software Research Associates, Inc., 2-32-8 Minami-Ikebukuro, Toshima, Tokyo, Japan
tkumazawa@acm.org

Abstract. We propose a recommender system which is based on a semisupervised classification algorithm designed for estimating users' preferred ranges of features. The system is targeted for new users, and it infers likings incrementally by presenting two alternatives to users in each step. To create the learning model, multidimensional scaling is employed to reduce the original feature space to a low-dimensional space. Then, the proposed classification algorithm, called geometrical exclusion, effectively finds a region which is not preferable for users and is to be excluded from the model in the reduced space. The algorithm consists of simple geometrical operations that are based on preference information obtained from users. An experiment by simulation is conducted to measure the performance of the system, and the result indicates that it can produce good recommendations with a practical number of user interaction steps. We also report statistics collected from our system deployed in a commercial web service.

Keywords: Recommender system · Interactive preference learning Multidimensional scaling · Geometrical operations

1 Introduction

Recommender systems have been intensively studied and attracted industry's interest, and now many web sites are equipped with recommender systems. They are usually classified into two categories: *content-based recommenders*, whose estimation process depends on similarity between items, and *collaborative*

M. Virvou et al. (Eds.): JCKBSE 2018, SIST 108, pp. 86–96, 2019.
https://doi.org/10.1007/978-3-319-97679-2_9

recommenders, which harness information gathered from people with similar preferences [1,17]. Collaborative recommendation has been demonstrated to be a powerful tool to estimate users' preferences, but to infer new users' preferences is intrinsically difficult. This is called *the cold start problem* [15].

In this paper, we propose a content-based recommender system for estimating new users' preferences to tackle the cold start problem. Preferences can be estimated by feature-wise interaction with the user as in critiquing systems [18], but it requires a high degree of user effort [13]. Considering the widespread use of mobile smart devices, such as smartphones and tablets, a complex user interface and/or an interface that needs substantial number of operations would be unacceptable. To invent a new sophisticated user interface could solve the problem [3], but to master a new interface would be a burden to casual users. To promote users' instant understanding, we opt to simply present two items to a user each turn and ask her to choose preferred one. Additionally, the proposed system can produce recommendations from item sets of various sizes.

The recommendation problem is commonly formulated as estimating *ratings* of items that have not been seen by a user in the literature [1]. We formulate it as an online semisupervised classification problem instead. The problem is intuitively decomposed as follows: (1) creating an initial model from the whole dataset whose objects are all unlabelled, (2) presenting a set of objects selected from the dataset to a user, where cardinality of the set could possibly be one but is two in our case, (3) receiving response from the user with labellings of objects, i.e., either *preferred* or *not preferred*, (4) updating the model based on the labellings, and (5) predicting labels of unlabelled objects by referencing the model. Precision of the prediction would increase through iterating the steps 2–4.

Our system uses a model embedded in a two-dimensional space as the initial model and employs multidimensional scaling (MDS) [5] to create it. MDS is a technique to construct a low-dimensional representation while preserving the distances between objects as well as possible. It has been mainly used as a visualisation technique to convert high-dimensional dataset to two- or three-dimensional image, as in [2]. Some recommender systems employ MDS or related visualisation techniques to create a two-dimensional map to be presented to a user, and user navigation is done in that map directly [6,9,10]. Giamattei and Sholtz employed correspondence analysis to visualise product recommendations and catalogues in a two-dimensional space [7]. In [16], MDS is used to embed both labelled and unlabelled objects in a reduced space, and linear discriminant analysis (LDA) is taken in the reduced space to classify objects. Khoshneshin and Street proposed a collaborative filtering algorithm that simultaneously embeds users and items in a reduced space by using MDS [11]. Le and Lauw proposed an embedding algorithm of both users and items based on a Bayesian approach, given preference data of items evaluated by users [12]. In [14], the problem of localising new users and items into an existing embedding is formulated as a quadratic programming. Grad-Gyenge et al. proposed a recommendation method based on a graph embedding [8]. To the best of our knowledge, there has been no attempt to propose an online classification algorithm which directly updates a low-dimensional model.

We present the proposed algorithm, which we call *geometrical exclusion*, used in our system in Sect. 2. Our algorithm has several number of unique features, such as fast learning inspired by binary search and regularisation of the learning process by referring to the original feature space to mitigate the degraded precision caused by restricting to learn in a two-dimensional space. In Sect. 3, we report the result of a simulation based experiment to measure the performance of our system with varying sizes of item sets. The experimental result shows that our algorithm is able to recommend favourable items with relatively steady and practical number of steps. Further, we deployed our system as a commercial web service as reported in Sect. 4. The result also indicates that our algorithm performs well in a production setting. We conclude our discussion in Sect. 5 with some notes on future work.

2 Geometrical Exclusion

In this section, we explain how to select objects to be labelled by a user and how the model is updated geometrically by using the information obtained from labelled objects and by excluding a region of the model.

2.1 Basic Idea

In our setting, the system presents a pair of objects to be labelled to a user. Assuming that there is no prior knowledge about user's liking, an efficient way is to select objects in a space along the line of binary search. We illustrate this by an example. Suppose we have a feature f_1 that ranges from 0 to 100. We want to know user's liking as a subrange, so we divide the range at the midpoint and present representative objects, which we call candidates, of each divided subrange to the user. Here we take candidates c_1 and c_2 from around 20 and 80 respectively since they are clearly distinguishable and not too extreme. If the user chooses c_1, we retain $[0, 50]$ and discard the rest. In the next turn, we divide the remaining range, present candidates around 10 and 40 to the user and discard a half according to the user's choice. Estimation proceeds in the same way until the size of the remaining range turns to be less than some threshold.

Unfortunately, the above estimation process of user's liking cannot be straightforwardly extended to two features in a two-dimensional space. The easiest way is to select on-axis candidates and to estimate the range feature-wise, but we consider a more general case where candidates are selected from an arbitrary axis pivoting at the centre. Suppose that there are two features, and we present two candidates to the user at each turn. In this case, however, we can see that a single turn is not sufficient to determine the right direction to proceed, because we cannot determine which of the features the user takes more seriously when she chooses one candidate. In other words, one feature of a candidate chosen by the user may not be preferable, but she chooses the candidate in favour of the other feature. So we have to take another axis that is orthogonal to the previous one and present another pair of candidates, augmenting the information about user's liking. If we extend the argument to n-dimensional case this way, the situation gets worse and more complicated since we need an exponential number of candidate pairs, i.e. 2^{n-1}, to infer user's liking.

Our solution to treat a high-dimensional feature space is to reduce its dimensionality. We employ classical MDS using Euclidean distance as the distance measure [5]. We focus our argument on reduction to a two-dimensional space in this paper, but the algorithms themselves could be applied to other cases.

To estimate original feature values from coordinates in the reduced space, we can employ linear regression. By calculating ranges of the original features from a survived region in the reduced space, we can predict whether a new object is *preferred* by a user. Regression is also applied to check whether a selected pair of candidates is appropriate to be labelled by a user because magnitude relation of a feature of the candidate pair does not necessarily coincide with the slope of a regression plane. Candidate pairs that have a feature whose magnitude relation does not coincide with a regression plane are not presented, and this regularises the learning process of user's liking as ranges in the original space.

2.2 Algorithm of Geometrical Exclusion

After reducing the dimensionality to two, we partition the search space so as to make it filled with tiles. Hereafter we can employ various kinds of operations applicable to a pixelated plane by treating a tile as a pixel, including elementary geometrical operations like drawing a line. The learning process is therefore easily and directly visualisable. We believe that this property of our algorithm makes learning process fairly tractable and parameter tuning relatively straightforward.

The algorithm of geometrical exclusion is shown in Fig. 1. We first calculate the mean m of the distribution of remaining data in the reduced space. Next, we find a main-axis a_1 so that the divergence along a_1 is the largest, by applying principal component analysis after setting m as the origin of the space, and then find a sub-axis a_2 which is orthogonal to a_1 and goes through m (line 5). For each axis a_i (in a_1 and a_2), we calculate the distribution of remaining data along a_i, then select pivot points p_1 and p_2 on a_i at 20 and 80 percentile respectively (line 7). We find candidates c_1 and c_2 so that they are the nearest to p_1 and p_2 respectively and their magnitude relations of all features coincide with the ones estimated from coordinates in the reduced space and regression planes for each feature (lines 8–10). We then display c_1 and c_2 to the user and ask her which one she prefers (line 11). Let r_i^p and r_i^n be the candidate points that are preferred and not preferred respectively (line 12). Then we determine a region to be excluded based on r_1^n and r_2^n, collect tiles in the region and exclude data on the tiles as follows. We set a boundary consisting of one line segment l_1 from r_1^n to r_2^n and two half-lines l_2 and l_3 (line 14). One half-line starts from r_1^n, taking the same direction as the direction from m to a pivot point from which r_1^n is selected. The other half-line is drawn likewise. We divide the space by the boundary (lines 15–18). Then we decide which region is *not* to be excluded based on the majority rule (line 19). The region which contains more points within

```
 1: Initialise search space S
 2: repeat
 3:   Fill S by colour k_1
 4:   Analyse S by principal component analysis
 5:   a_{1,2} ← axes corresponding to two eigenvalues
 6:   for i ← 1, 2 do
 7:     Select pivot points p_{1,2} from axis a_i
 8:     repeat
 9:       Select candidates c_{1,2} around p_{1,2} respectively
10:     until coincide?(S, c_{1,2})
11:     Display candidates c_{1,2} and wait response
12:     r_i^p ← candidate chosen, r_i^n ← candidate not chosen
13:   end for
14:   Place line segments l_{1,2,3} based on the position of r_{1,2}^n
15:   Draw l_{1,2,3} on S by colour k_3
16:   Select an arbitrary tile t not belonging to l_{1,2,3}
17:   Flood-fill S from t by colour k_2
18:   k ← colours of r_{1,2}^p and the mean
19:   u ← subscript of the majority of colours in k
20:   Exclude k_{3-u} and k_3 coloured tiles from S
21: until converged?(S)
```

Fig. 1. Geometrical exclusion algorithm.

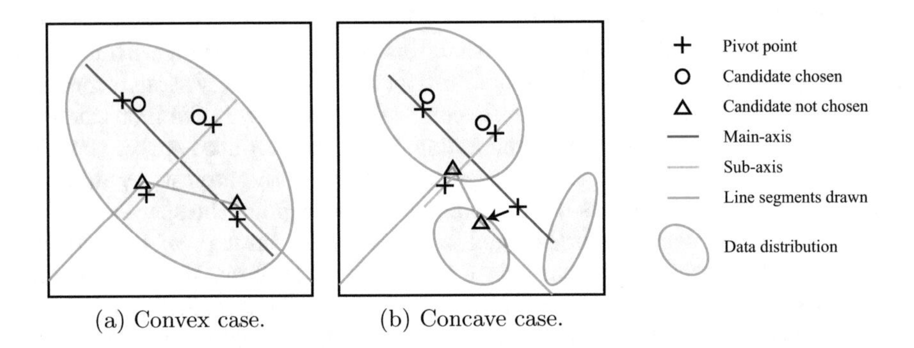

(a) Convex case. (b) Concave case.

Fig. 2. Schematic diagrams of geometrical exclusion.

r_1^p, r_2^p and m than the other is to be remained. Complement of the majority region (i.e. the boundary and the minority region) is to be excluded (line 20).

Our strategy to determine the excluded region is to collect tiles in the region by colouring them like drawing a picture. The shape of the excluded region may be concave (see Fig. 2), so its formal analysis is difficult because it needs careful treatment of exceptions. Instead, we simply draw a boundary and fill a region isolated by the boundary. We employ Bresenham's algorithm [4] for the former, and a classic flood fill algorithm for the latter respectively.

Table 1. Correlation matrices used in the experiment.

(1) Rental residences.

F.1	F.2	F.3	F.4
1.0	-	-	-
0.8	1.0	-	-
-0.2	0	1.0	-
-0.1	0.3	0.0	1.0

(2) Three separated groups.

F.1	F.2	F.3	F.4
1.0	-	-	-
0.9	1.0	-	-
0	0	1.0	-
0	0	0	1.0

(3) Three groups.

F.1	F.2	F.3	F.4
1.0	-	-	-
0.9	1.0	-	-
-0.1	0.3	1.0	-
0.2	-0.2	0.1	1.0

(4) Two separated groups.

F.1	F.2	F.3	F.4
1.0	-	-	-
0.85	1.0	-	-
0	0	1.0	-
0	0	-0.7	1.0

3 Experimental Evaluation by Simulation

We conducted an experiment to evaluate the performance of our algorithm, with artificially generated datasets and a user behaviour simulator. The reason to rely on a simulation is that we need to compare the resulting performance to those of other systems, measured under the same experimental setting and user inputs.

The datasets consists of randomly generated objects, each of whom has four features. Each feature is real-valued and sampled from the standard normal distribution. We prepared several number of datasets with differing sizes and correlations of features. Correlation between each feature is one of the key factors of our algorithm's performance, so we tested against four configurations of feature correlations, as shown in Table 1. *Rental residences* configuration is reproduced from the correlation matrix of the actual rental residences dataset, and the other three configurations are artificially chosen for comparison.

The design of the simulator roughly follows a way employed in [13]. In each trial, the simulator randomly selects an object from the dataset at first, and it is used as a *target object* during the trial. Additionally, a set consisting of objects similar to the target object, including the target object itself, found within the dataset is constructed, and we call it the *target set* of the trial. Members of the target set are objects whose every feature value falls within a certain range, centred at values of the target object. We tested several number of width configurations of the range, relative to the standard deviation σ of the feature values, from $\pm 0.2\sigma$ to $\pm 1.0\sigma$. Next, the simulator runs an algorithm to be tested and get two objects recommended by the algorithm. If at least one of the recommended objects matched an object in the target set, the trial finishes successfully. When both objects failed to match, the simulator calculates the Euclidean distance between the target object and each recommended object and takes the closer one to the target object. The object chosen above is passed to the algorithm as if it were a user input. Then the algorithm produces next recommendation, and the simulation iterates this way automatically.

To make an automatic simulation possible, without relying upon actual users' behavioural data, we employed an experimental setting that reflects reality in some degree, though significantly simplified. A user can compare two items based on Euclidean distance to the target item, and she chooses a closer one as her preferred item. This is based on assumptions that users have clear figures of ideal items they want to buy, and also they can precisely measure the distance between

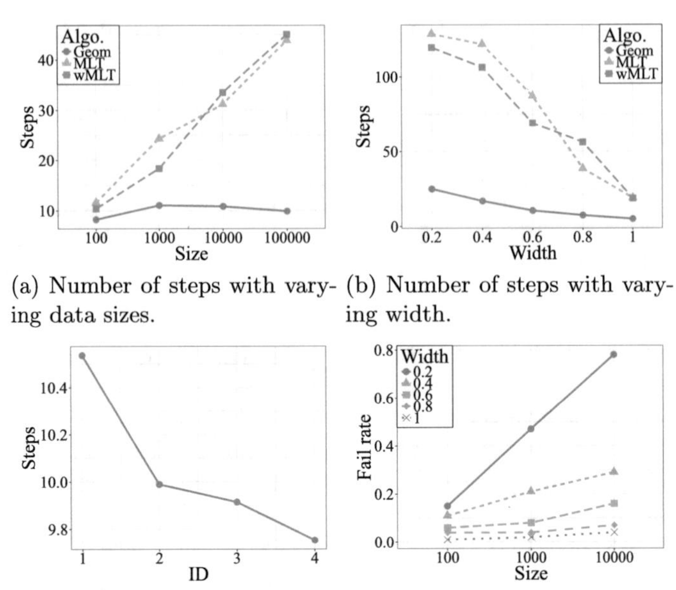

(a) Number of steps with varying data sizes.

(b) Number of steps with varying width.

(c) Number of steps with varying correlation coefficients.

(d) Fail rate with varying data sizes and width.

Fig. 3. Average number of steps and fail rate. Default data size, width and correlation configuration are 10,000, 0.6 and *rental residences* resp.

Table 2. Average sizes of target sets. Data size is 10,000 and correlation configuration is *rental residences.*

Width	$\pm 0.2\sigma$	$\pm 0.4\sigma$	$\pm 0.6\sigma$	$\pm 0.8\sigma$	$\pm 1.0\sigma$
Target set size	4.9	52.5	218.1	559.6	1099.7

an ideal item and a presented one, comparing each feature independently. This may seem too unrealistic, but we want to show basic characteristics and relative advantage of our algorithm here, in a clear setting involving few confounding factors, ensured by such strong assumptions.

We compared the performance of three algorithms with slight modifications: *Geometrical Exclusion*, *More Like This (MLT)* [13], *Weighted More Like This (wMLT)* [13]. In order to make wMLT work, we had to discretise values of features before comparison. We are aware of potential influences that the way of this discretisation may have, but we simply divide the value range of each feature into 0.2σ width ranges in this experiment, to keep the number of parameters we need to examine as few as possible. Another difference to McGinty's setting is that the initial pair of objects to be presented for MLT and wMLT are selected randomly in our experiment.

We ran the simulation with varying data sizes, width of the target set range and correlation configurations. Results are averaged over 100 runs for each setting and are shown in Fig. 3 and Table 2. Figure 3(a) clearly indicates that our algorithm is able to produce favourable recommendations within the range of practical number of steps under a realistic setting. The number of steps necessary to recommend a preferred item varies steadily around 10 for all size settings. This contrasts with that of the MLT and wMLT algorithms, whose number of steps increases logarithmically as the data size increases. The relative target set size for the case of 10,000 objects, $\pm 0.6\sigma$ width and *rental residences* correlation configuration is about 2%, as shown in Table 2, and this is a moderate number for our service. Standard deviation of the steps under the above setting is 8.57 for our algorithm, whereas that of MLT and wMLT are 18.87 and 43.16 respectively. Figure 3(b) shows that the number of steps decreases when the target set size increases for all three algorithms, and this conforms to our intuition. Our algorithm outperformed MLT and wMLT in all five width settings. Correlation configurations are arranged from the most difficult to recommend, *rental residences*, to the easiest, *two separated groups*. The number of steps necessary for our algorithm, shown in Fig. 3(c), decently reflects those relative difficulties.

Our algorithm occasionally fails to recommend a preferred item because its tile-wise exclusion mechanism may drop all the tiles that contain target objects before presenting them to the user. In Fig. 3(d), fail rate statistics with varying data sizes and width are shown. Fail rates for the range of width broader than or equal to $\pm 0.6\sigma$ are kept fairly low for all data sizes.

4 Deployment as a Web Service

We implemented the proposed algorithm as an online recommender system embedded in a commercial web service to search rental residences, and we made our service publicly accessible during five weeks. During this period, we collected log data to record user behaviour, from which we obtained statistics shown below.

The rental residence dataset of our system has four integer-valued features: monthly rent, floor area, age and distance from the nearby station. These features are selected out of several tens of available features, considering suggestions by domain experts. The dataset is grouped by nearby train/subway station. The number of groups is several thousand, and the number of residences contained in each group ranges from one to nearly ten thousand.

A user of our service is asked to choose a station first. Then the system loads the dataset and creates an initial model specific to that station. After the setup, two residences are displayed and the user is asked to choose more preferable one (see Fig. 4). The learning step progresses in accordance with the algorithm of the geometrical exclusion, however, to enhance usability, we made a modification so that the steps cannot be repeated more than four times. After the learning steps, the system computes ranges of the features, which represent user's likings for rental residences. Finally, residences classified as *preferred* are displayed.

As a result of the publicly opened deployment, the total number of sessions was 6,318, though, the number of sessions with user activity was 2,963 (46.9%), meaning that more than half of the sessions finished with no user responses.

Fig. 4. A screenshot of the prototype displaying 1. floor plans and 2. features of two residences. When a user presses 3. one of the buttons at the bottom which is under a preferable residence, 4. number of operations left is decremented and new residences are shown. Several parts containing sensitive information are blurred.

Meanwhile, 1,970 (31.2%) sessions were *completed*, that is, substantial amount of users finished the learning process and got final recommendations. The number of unique users who completed their sessions was 1,455, which was 38.1% of the total number of unique users, 3,815. The average number of sessions per user was 1.83, and its standard deviation was 8.63. Average time consumed to finish the process was 144 s, and its standard deviation was 147. Considering the above rate of completion, this falls within a tolerable range of ordinary users.

5 Conclusion and Future Work

We presented a novel recommender system that directly utilises a two-dimensional space, reduced from the feature space, to embed a learning model for semisupervised classification. The proposed learning algorithm is based on geometrical operations on the reduced space. By a simulated experiment, we showed that our system is able to estimate preferred ranges of features effectively. The number of steps necessary to complete the estimation is kept within a practical range and stable enough with varying sizes of item sets. We believe that this property made our system amusing for many users involved in an open trial, and thus, it resulted in high completion rate.

As future work, we plan to formally analyse properties of our algorithm such as convergence of the learning process by geometrical exclusion. In addition, it is desirable to measure our system's performance in a "real" setting rather than in a simulated experiment. Ideally, recommender systems' performance should be compared in a practical production environment in an A/B test fashion.

References

1. Adomavicius, G., Tuzhilin, A.: Toward the next generation of recommender systems: a survey of the state-of-the-art and possible extensions. IEEE Trans. Knowl. Data Eng. **17**(6), 734–749 (2005)
2. Basalaj, W.: Incremental multidimensional scaling method for database visualization. In: Proceedings of Visual Data Exploration and Analysis VI, SPIE, vol. 3643, pp. 149–158 (1999)
3. Baur, D., Boring, S., Butz, A.: Rush: repeated recommendations on mobile devices. In: Proceedings of the 15th International Conference on Intelligent User Interfaces (IUI 2010), pp. 91–100. ACM (2010)
4. Bresenham, J.E.: Algorithm for computer control of a digital plotter. IBM Syst. J. **4**(1), 25–30 (1965)
5. Cox, T.F., Cox, M.: Multidimensional Scaling, 2nd edn. Chapman and Hall/CRC, Boca Raton (2000)
6. Frank, J., Lidy, T., Hlavac, P., Rauber, A.: Map-based music interfaces for mobile devices. In: Proceedings of the 16th ACM International Conference on Multimedia (MM 2008), pp. 981–982. ACM (2008)
7. Giamattei, M., Scholz, M.: Exploiting correspondence analysis to visualize product spaces. In: Proceedings of the 7th Conference of the Italian Chapter of AIS (2010)
8. Grad-Gyenge, L., Kiss, A., Filzmoser, P.: Graph embedding based recommendation techniques on the knowledge graph. In: Adjunct Publication of the 25th Conference on User Modeling, Adaptation and Personalization, pp. 354–359 (2017)
9. Kagie, M., van Wezel, M., Groenen, P.: Map Based Visualization of Product Catalogs. ERIM Report Series Research in Management ERS-2009-010-MKT, Erasmus Research Institute of Management (ERIM) (2009)
10. Kagie, M., van Wezel, M., Groenen, P.J.F.: A graphical shopping interface based on product attributes. Decis. Support Syst. **46**(1), 265–276 (2008)
11. Khoshneshin, M., Street, W.N.: Collaborative filtering via euclidean embedding. In: Proceedings of the Fourth ACM Conference on Recommender Systems (RecSys 2010), pp. 87–94. ACM (2010)
12. Le, D.D., Lauw, H.W.: Euclidean co-embedding of ordinal data for multi-type visualization. In: Proceedings of the 2016 SIAM International Conference on Data Mining, pp. 396–404 (2016)
13. McGinty, L., Smyth, B.: Comparison-based recommendation. In: Proceedings of the 6th European Conference on Advances in Case-Based Reasoning (ECCBR 2002), pp. 575–589. Springer (2002)
14. O'Shaughnessy, M.R., Davenport, M.A.: Localizing users and items from paired comparisons. In: 26th IEEE International Workshop on Machine Learning for Signal Processing (MLSP 2016), pp. 1–6 (2016)
15. Schein, A.I., Popescul, A., Ungar, L.H., Pennock, D.M.: Methods and metrics for cold-start recommendations. In: Proceedings of the 25th Annual International ACM SIGIR Conference on Research and Development in Information Retrieval (SIGIR 2002), pp. 253–260. ACM (2002)
16. Trosset, M.W., Priebe, C.E., Park, Y., Miller, M.I.: Semisupervised learning from dissimilarity data. Comput. Stat. Data Anal. **52**(10), 4643–4657 (2008)

17. Yang, X., Guo, Y., Liu, Y., Steck, H.: A survey of collaborative filtering based social recommender systems. Comput. Commun. **41**, 1–10 (2014)
18. Zhang, J., Pu, P.: A comparative study of compound critique generation in conversational recommender systems. In: Proceedings of the 4th International Conference on Adaptive Hypermedia and Adaptive Web-Based Systems (AH 2006), pp. 234–243. Springer (2006)

Quantitative Evaluation of IT Management Activity Knowledge

Shuicihiro Yamamoto(✉) and Shuji Morisaki

Nagoya University, Furo-cho, Chikusa-ku, Nagoya, Aichi 464-8601, Japan
syamamoto@acm.org, morisaki@is.nagoya-u.ac.jp

Abstract. It is well known that IT management is the critical success factor for operating IT systems. However, the quantitative knowledge to evaluate the IT management capability of organizations has not been known. This paper proposes a quantitative IT management activity knowledge to evaluate IT management capability of organizations. The IT management activity Knowledge (ITMAK) is defined based on the 8 composite activities, that are IT management basic, IT risk analysis, vision construction, communication, product design, process design, investment optimization, and human resource development. The proposed ITMAK has also been applied to Japanese IT management experts. The result shows that ITMAK can effectively be applied to visualize the IT management activities of the organizations.

Keywords: IT management · Activity knowledge · Visualization
Quantitative evaluation · Case study

1 Introduction

It is well noticed that IT management is the important success factor to achieve business values using IT. There were many approaches, methods, and guidelines so far. To provide IT management guidelines for enterprises, it is necessary to assess the organizational knowledge on IT management activities. The reason is that effective organizational knowledge is necessary to achieve good IT management activities. Therefore, it is necessary to evaluate organizational knowledge to manage IT. For example, it is difficult to achieve appropriate IT management activities based on ill-defined IT management process and artifact configuration without IT management vision. The knowledge translation from generic IT management methods to the specific organizational knowledge of IT management activities is necessary. If the knowledge translation is missing, the education of generic IT management methods is useless.

The research objective of this paper is to provide a new lightweight method to visualize IT management activity knowledge.

The rest of the paper is structured as follows. Section 2 describes related work. The IT Management Activity Knowledge (ITMAK) model is proposed in Sect. 3. Evaluation of the proposed ITMAK is described in Sect. 4. Discussions are shown in Sect. 5. Section 6 summarizes this paper.

M. Virvou et al. (Eds.): JCKBSE 2018, SIST 108, pp. 97–107, 2019.
https://doi.org/10.1007/978-3-319-97679-2_10

2 Related Work

Arthur [1] proposed the quality metric on the development document maintainability. SW-CMM [2] showed Requirements management as the Key Process Area (KPA) in software process capability aspects. CMMI-SE/SW adds Requirements Development as the KPA. Dautovic [5] developed a tool to automatically evaluate software development document described by word processor based on the practical review rules.

Wiegers [3] described 20 questions to diagnose requirements practices. Each question shows four choices to select the current levels of organizational practice. The answers of the questions can be effectively used to improve requirements techniques of organizations. These questions are bounded to technical aspects of requirements. It was not designed to measure organizational capability of requirements specification.

Pohl [4] describes a capability model to validate requirements with three levels. The three levels of requirements validation are minimal, standard, and advanced validation. These levels are defined through five validation aspects of activities. The aspects are context consideration, artefacts-context dimension, artefacts-documentation dimension, artefact-agreement dimension, and activity execution. Each aspect has sub activities. The number of sub activities of five aspects are 3, 8, 4, 4, and 5 respectively. The minimal validation level defines activities that each project should consider. The standard validation level defines activities that projects should achieve. The advanced validation level defines activities that quality critical project should achieve.

IT management maturity models have been proposed, such as ISO/IEC 15504 process assessment [13], CMMI (Capability Maturity Model Integration) [14], Business Process Maturity Model (BPMM) [15], IT-CMF [16]. IT governance maturity models, such as COBIT IT BSC [17], CGEIT [18], IT project Auditing Assurance framework [19], have also been proposed. Proença and others [20] pointed that these assessment methods are heavy to measure organizational capability.

IT management improvement approach based on capability index was introduced by Yamamoto [6]. The index was elicited based on IT Infrastructure Library (ITIL) [21–25] by analyzing human activity [7]. Requirements Specification Capability Index (RSCI) was also proposed to evaluate organizational capability to visualize organizational requirements specification capability [8].

This paper proposes a lightweight method to visualize IT management activities based on the approach proposed by RSCI.

3 Design of IT Management Knowledge Model

To design IT management activity knowledge, basic Meta model of activities are defined and then knowledge to support activities is elicited for activities. There are primitive and composite activities. Composite activities contain multiple primitive activities. Activities are supported by corresponding knowledge. When people act something, there may be knowledge how to act. Knowledge supports people do the right activity. Based on the consideration, the Meta model of activities can be defined as shown in Fig. 1.

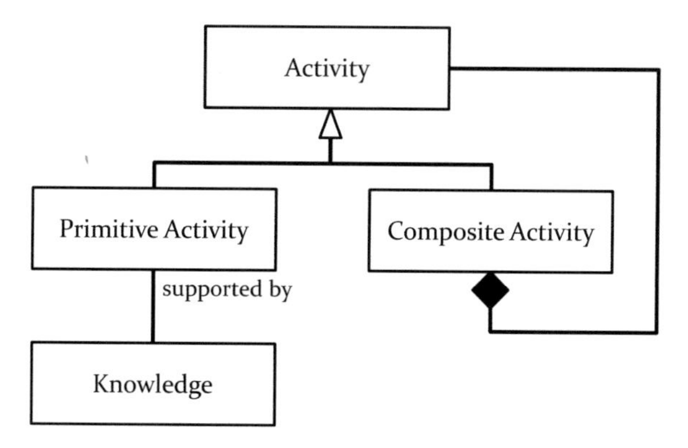

Fig. 1. Meta model of activities.

Each activity knowledge is evaluated based on the levels shown in Table 1. Knowledge levels are assigned to knowledge types. The five knowledge types in Table 1 are none, tacit, intermediary, explicit in local, and explicit in global. The intermediary knowledge has been introduced by Yamamoto and others [9–12] to explain knowledge sharing process in computer mediated communication such as social networking services. Intermediary knowledge is fragmented but described using some key words. Memo and unstructured note are typical examples of intermediary knowledge. Explicit knowledge is generalized and authorized by enterprises. The difference between explicit in local and global knowledge is the range of divisions. Explicit knowledge in local is shared in limited divisions of enterprises, whereas explicit knowledge in global is shared in the whole enterprise.

Table 1. Levels of Activity Knowledge.

Level	Knowledge type	Explanation
0	None	Activity is not carried out
1	Tacit	Activity is defined orally without prescriptions
2	Intermediary	Activity is defined by fragmented keywords
3	Explicit in department	Activity is defined by department manual
4	Explicit in enterprise	Activity is defined by enterprise standard manual

In this paper, composite activities for IT management are defined by basic, risk analysis, vision construction, communication, process design, product design, investment optimization, and human resource design as shown in Table 2. Table 2 also shows the 49 IT management primitive activity knowledge.

Table 2. Configuration of IT Management Activity Knowledge.

Composite activity	Primitive activity knowledge
IT management basic (5)	(1) Basics of IT management are clarified
	(2) The needs of IT management are clarified
	(3) The effects of IT management are clarified
	(4) The needs of IT strategy are clarified
	(5) The effects of the latest IT technology are clarified
IT risk analysis (7)	(1) IT risk management principles are defined
	(2) IT risk management plan is defined
	(3) IT risk management procedure is defined
	(4) IT risk management information is defined
	(5) IT risk is evaluated
	(6) IT incident information sharing is defined
	(7) IT risk mitigation procedure is defined
IT management vision construction (9)	(1) Role of IT management is defined to achieve IT strategy
	(2) IT management organization is developed
	(3) The priority of IT management investment is clarified
	(4) IT management in development phase is clarified
	(5) Roles of IT management department is clarified
	(6) IT management department role for development team is clarified
	(7) Responsibility of development team is clarified for IT management
	(8) IT strategy promotion team is clarified
	(9) The need of Alignment between business and IT is clarified
IT management communication (7)	(1) IT management roles in development is shared
	(2) Utilization policy of IT management in development is shared
	(3) The goals of IT management is shared with development team
	(4) The change caused by IT management is realized by development team
	(5) Problem solving process to utilize IT management is clarified between IT management and development teams
	(6) Sharing method of IT management practices is clarified
	(7) IT management ROI is clarified among top management, IT management, and development teams
IT management product design (5)	(1) Existing artifacts are adopted to IT management
	(2) IT management artifacts are clarified
	(3) Utilization policy of IT management artifacts is clarified
	(4) Collaborative artifacts with external partners are clarified

(*continued*)

Table 2. (*continued*)

Composite activity	Primitive activity knowledge
	(5) Reduction of redundant IT management artifacts is clarified
IT management process design (5)	(1) Base line development process is clarified
	(2) Future development process is clarified
	(3) Utilization policy of IT management process is clarified
	(4) Collaborative processes with external partners are clarified
	(5) Reduction of redundant IT management processes is clarified
IT management investment optimization (6)	(1) Cost of IT management implementation is clarified
	(2) IT management cost and merit are verified beforehand
	(3) IT management conformance is considered for enterprise optimization
	(4) Compatibility to enterprise optimization is evaluated in IT management implementation
	(5) Utilization and effectiveness of IT management are measured after IT management implementation
	(6) Issues of IT management utilization are analyzed and resolved
IT management Human Resource Development (5)	(1) Human resource is developed to propose improved IT management process
	(2) Human resource who are familiar with IT management as well as development is assigned to business management level
	(3) Opportunity to learn business management knowledge is provided for IT management human resources
	(4) Opportunity to understand development process is provided to IT management human resources
	(5) IT management utilization skill seminar is provided to IT engineers

The ITMAK (IT Management Activity Knowledge) consists of eight composite activities. The primitive activities of a composite activity are derived to achieve the composite activity. For example, IT risk management principles, plan, procedures and information are necessary to be defined to achieve IT risk analysis activity. IT risk is necessary to be evaluated as the outcome of IT risk analysis. IT incident information sharing and IT risk mitigation procedure activities are also necessary to be defined as the result of IT risk analysis.

The ITMAK value V(x) is defined as the mean value of (level a_i)/4 for all primitive activity a_i of the composite activity x. The value 4 is the maximum level of knowledge as shown in Table 1. For example, IT management vision construction includes 5 primitive activities. Suppose the levels of five activities are 1, 2, 1, 3, and 1. Then V (IT management vision construction) is calculated as $((1 + 2 + 1 + 3 + 1)/4)/5 = 2/5 = 0.4$. The meaning of the value 0.4 is more than those of oral tacit and less than intermediary knowledge, because $0.4 \times 4 = 1.6$. If the value is greater than 2, there is the evidence more than the intermediary knowledge on the activity.

4 Evaluation of the IT Management Activity Knowledge

The evaluation of the proposed ITMAK has been evaluated by Japanese IT management experts as shown below.

4.1 Subjects

The ITMAK questionnaire form was provided to 14 experts. The ITMAK questionnaire consists of 49 questions corresponding to IT management activities defined in Table 2. There were 11 client company experts and 3 independent consultants.

4.2 Evaluation Result

The result of the questionnaire for 14 subjects is shown in Table 3 by discriminating eight composite activities of RSCI. The value is defined as the summation of values for each activity. The time to answer the questions is approximately 10 min.

Table 3. Application result of requirements specification capability index.

	Basic	Risk analysis	Vision definition	Communication	Product design	Process design	Investment	HRD
Client	0.677	0.604	0.551	0.399	0.409	0.464	0.496	0.5
Consul	0.5	0.488	0.435	0.298	0.267	0.4	0.5	0.35
Mean	0.639	0.579	0.526	0.378	0.379	0.425	0.470	0.361

The mean value of client is 0.512, whereas those value of consultants is 0.405. This showed that clients have more knowledge than intermediary level, although consultants have less knowledge than intermediary level. The ITMAK value of client companies is higher than those of consultants except for investment activities. The ITMAK value of consultants is higher than those of clients only for the investment activity.

The correlation between ITMAK is shown in Table 4. The number with under-lines shows relatively high correlation. Figures 2, 3, 4, and 5 show the relationship between Communication and Product design, the relationship between Basic and IT risk analysis activity knowledge, the relationship between Vision construction and Investment activity knowledge and the relationship between Vision construction and Product design activity knowledge, respectively.

Table 4. Correlation between IT management activity knowledge.

	Risk analysis	Vision definition	Communication	Product design	Process design	Investment	HRD
Basic	0.715	0.2078	0.4586	0.1801	0.3487	0.1729	0.1243
Risk analysis		0.4165	0.3135	0.2376	0.4549	0.3669	0.0718
Vision definition			0.5908	0.6039	0.4932	0.6535	0.0841
Communication				0.7235	0.5047	0.4275	0.0587
Product design					0.4754	0.5056	0.0339
Process design						0.3973	0.0713
Investment							0.2755

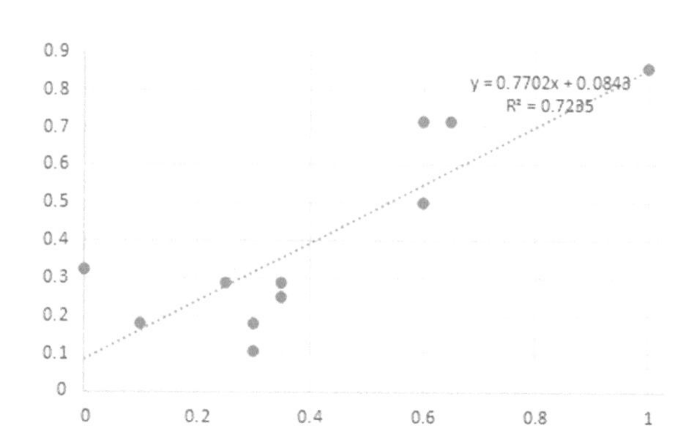

Fig. 2. Relationship between Communication and Product design activity knowledge.

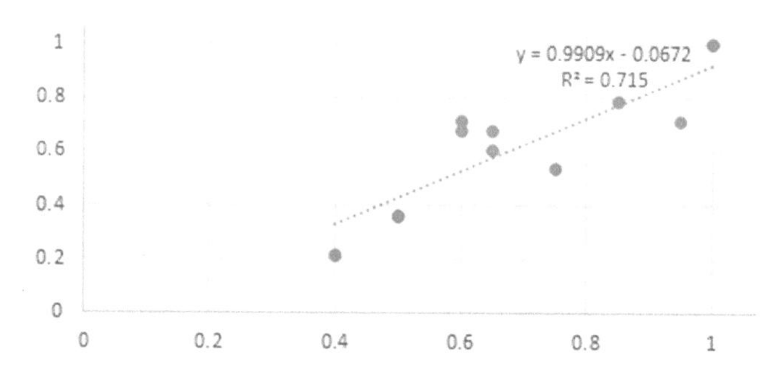

Fig. 3. Relationship between Basic and IT risk analysis activity knowledge.

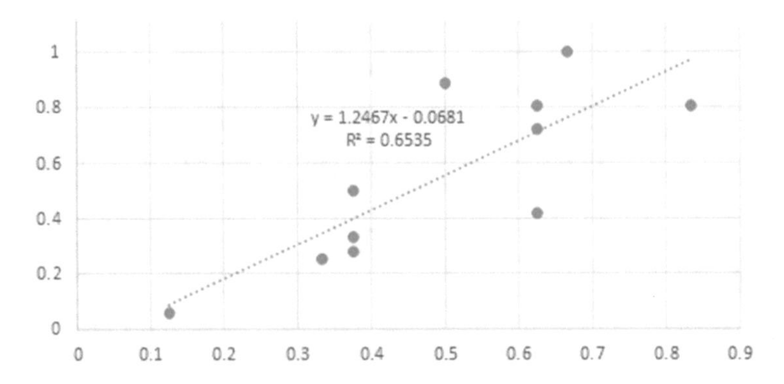

Fig. 4. Relationship between Vision construction and Investment activity knowledge.

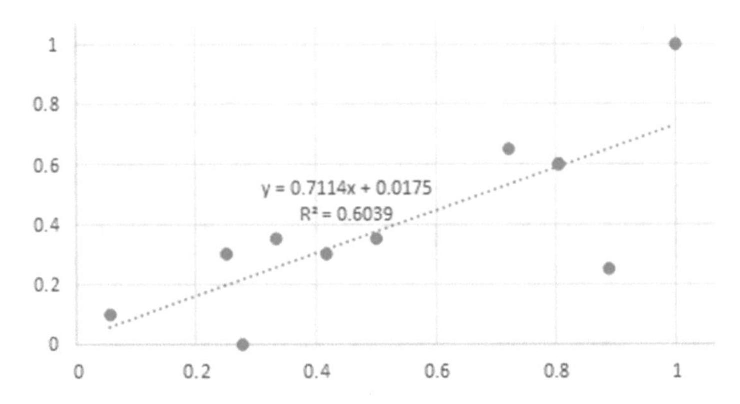

Fig. 5. Relationship between Vision construction and Product design activity knowledge.

5 Discussion

5.1 Effectiveness

As mentioned above, ITMAK is easily applicable to detect quantitatively week points of organizational knowledge and improve IT management knowledge by taking measures for the detected week points of knowledge. The time to answer 49 questions was very low. This shows the efficiency of the proposed method. To realize effective IT management, IT management knowledge on vision construction, communication, product design, process design, investment and HRM is necessary as shown in Table 3.

The mean ITMAK value on product and process are 0.38 and 0.43 respectively. This showed that existing IT management methods have still not been well established for practical IT management experts. In other words, the proposed ITMAK is useful to show the limitation of current IT management approaches quantitatively. This also disclosed that new practical IT management method is necessary to fit field experts. These arguments showed the effectiveness of ITMAK to explain the necessity of the new approaches to resolve issues on organizational IT management activities quantitatively.

Moreover, ITMAK can quantitatively visualize IT management knowledge inter organizations. Industry wide IT management knowledge can also be visualized. For example, IT management knowledge of each industry will be clarified. And then, each company can evaluate the organizational knowledge on IT management according to the industry knowledge criteria. For example, standard values of ITMAK can be provided for each industry. Organizations can objectively recognize the IT management activities according to the industry standard ITMAK. If ITMAK value of an organization is lower than the standard value, IT management knowledge improvement shall be necessary. Client companies can rationally select vendors that have appropriate ITMAK value by providing necessary levels of ITMAK. If vendors have insufficient IT management knowledge, it is impossible to depend on vendors. The ITMAK can be applied to support for building consensus by providing appropriate level of IT management knowledge.

5.2 Application Method of ITMAK

ITMAK will help elicit improvement points by visualizing issues of IT management activities, because ITMAK defines knowledge evaluation index on IT management activities.

(1) ITMAK is useful to know if IT management is able to success by identifying necessary level of knowledge on IT management and investigating the corresponding value of the organizational knowledge.
(2) ITMAK is useful for identifying activities necessary to improve by investigating knowledge levels of activities. For example, cross-sectional evaluation of IT management can be achieved by company hearing based on ITMAC.

5.3 Comparison with Other IT Management Evaluation Methods

As shown in Sect. 4.2, the time to answer the 49 questions was approximately 10 min. This shows the efficiency of applying ITMAK. Other IT management evaluation methods, such as IT-CMF, need several months to evaluate organizational IT management capability.

ITMAK can also be applied to improve IT management activity of organizations. ITMAK clarifies weak points of IT management capability by identifying primitive activities of low index values. These low valued primitive activities are candidate activities to be improved.

5.4 Limitation

The purpose of this paper is to reveal that IT management knowledge level can quantitatively be evaluated through ITMAK. Therefore, the completeness of ITMAK has not been confirmed. The more complete ITMAK will be developed by more number of case studies. However, there may be another problem if the number of primitive and composite activities increases. The efficiency of assessing ITMAK decreases so that the number of ITMAK activities increases.

6 Conclusion

In this paper, the IT Management Activity Knowledge, ITMAK, for evaluating organizational IT management knowledge is proposed. The application result of ITMAK has also been clarified for Japanese IT management experts. Previous introduction of IT management approaches tried to achieve IT management by educating general IT management methods for members. However, it is more efficient to specify appropriate IT management methods based on the result by evaluating ITMAK of each organization. If the organization did not have sufficient IT management activity knowledge, it is useless to apply new methods and principles for the organization. The IT management approaches shall be fit to the organizational knowledge of IT management. ITMAK can visualize quantitatively the organizational knowledge of IT management. ITMAK provides a way to translate IT management approaches into the organizational knowledge.

References

1. Arthur, J., Stevens, K.: Assessing the adequacy of documentation through document quality indicators. In: Proceedings of the International Conference on Software Maintenance (1989)
2. CMU SEI, CMMI for Systems Engineering/Software Engineering, Technical Report CMU/SEI-2000-TR-018 (2000)
3. Wiegers, K.: Software Requirements, Practical techniques for gathering and managing requirements throughout the product development cycle. Microsoft (2003)
4. Pohl, K.: Requirements Engineering – Fundamentals, Principles, and Techniques. Springer, Heidelberg (2010)
5. Dautovic, A., Plösch, V, Saft, M.: Automatic checking of quality best practices in software development documents. In: 2011 11th International Conference on Quality Software, pp. 208–217 (2011)
6. Yamamoto, S.: A continuous approach to improve IT management. In: Procedia Computer Science, vol. 121. CENTERIS 2017, pp. 27–35 (2017)
7. Yamamoto, S.: A human activity based operational knowledge elicitation method. In: Procedia Computer Science, vol. 96. KES 2016, pp. 848–858 (2016)
8. Yamamoto, S.: An evaluation of requirements specification capability index. In: Procedia Computer Science, vol. 112. KES 2017, pp. 998–1006 (2017)
9. Yamamoto, S., Kanbe, M.: Knowledge creation by enterprise SNS. Int. J. Knowl. Cult. Change Manag. **8**(1), 255–264 (2008)
10. Kanbe, M., Yamamoto, S.: An analysis of computer mediated communication patterns. Int. J. Knowl. Cult. Change Manag. **9**(3), 35–47 (2009)
11. Kanbe, M., Yamamoto, S., Ohta, T.: A proposal of TIE model for communication in software development process. In: Nakakoji, K., Murakami, Y., McCready, E. (eds.) JSAI-isAI. LNAI, vol. 6284, pp. 104–115. Springer, Heidelberg (2010)
12. Masuda, Y., Shirasaka, S., Yamamoto, S., Hardjono, T.: The proposal of GDTC model for global communication on enterprise portal. In: Procedia Computer Science, vol. 112. KES 2017, pp. 998–1006 (2017)
13. ISO/IEC Std. 15504:2004: Information technology – Process assessment (2004)
14. CMU/SEI-2010-TR-034: CMMI for Services, V1.3 (2010)
15. Open Management Group: Business Process Maturity Model (BPMM) - Version 1.0, (2008)

16. Innovation Value Institute: The IT-CMF Framework. http://ivi.nuim.ie/it-cmf
17. Information Systems Audit and Control Association (ISACA), Control Objectives and Related Technology (COBIT). http://www.isaca.org/COBIT/
18. ISACA, CGEIT Review Manual, 7th Edition (2015)
19. Mkoba, E., Marnewick, C.: IT project success: a conceptual framework for IT project auditing assurance. In: SAICSIT 2016 (2016). https://doi.org/10.1145/2987491.2987495
20. Proença, D., Borbinha, J.: Maturity models for information systems - a state of the art. In: Conference on Enterprise Information Systems, pp. 1042–1049 (2016)
21. Hunnebeck, L.: ITIL Service Design, 2nd edn. The Stationery Office (2011)
22. Cannon, D.: ITIL Service Strategy, 2nd edn. The Stationery Office (2011)
23. Steinberg, L.A.: ITIL Service Operation (Best Management Practices), 2nd edn. The Stationery Office (2011)
24. Lloyd, V.: ITIL Continual Service Improvement (Best Management Practices), 2nd edn. The Stationery Office (2011)
25. Rance, S.: ITIL Service Transition (Best Management Practices), 2nd edn. The Stationery Office (2011)

Bibliometrics EEG Metrics Associations and Connections Between Learning Disabilities and the Human Brain Activity

Vasileios Stefanidis[1], Marios Poulos[2(✉)], and Sozon Papavlasopoulos[2]

[1] University of Ioannina, Ioannina, Greece
vstefan@cc.uoi.gr
[2] Ionian University Corfu, Corfu, Greece
{mpoulos,sozon}@ionio.gr

Abstract. In this paper we investigate the bibliometric association and connection between Electroencephalography (EEG) metrics of human brain and learning difficulties and disabilities. In EEG metrics included various metrics used from scientists in order to map the brain activity. In various big data bases as the biomedical databases there is an increasingly amount of data stored in, due to the breakthroughs in biology and bioinformatics. In recent years biomedical information is in the center of research growing the amount of data and thus making efficient information retrieval from scientists more and more challenging. Scientists, needs new tools and applications in order to extract scientifically important data based on experimental results and information provided by papers and journals. In this paper we are going to investigate methods in order to find connections between learning difficulties and disabilities and the brain operation and signaling.

1 Introduction

Electroencephalography (EEG) is a biometric method that can give you fairly detailed information on ongoing brain activity associated with perception, cognition and emotion.

There are various EEG-based Metrics we can find in literature as cognitive state for engagement and distraction as well as a mental workload metric [17], the memory workload, sort memory capacity, various EEG entropies, power spectra (PS), coherence (ITC – InterTrial Coherence) [19], the pre-stimulus noise (PSN), signal-to-noise ratio (SNR) and EEG amplitude variance across the P300 event window (CVERP) [18], the EEG rhythms and event related potentials (ERPs) [20] etc.

The metrics enable researchers to observe the unobservable brain, and incorporate a useful and reliable alternative to analyzing raw EEG data.

Following the revolution in biology and bioinformatics, biomedical information is in the center of research growing the amount of data and thus making efficient information retrieval from scientists more and more challenging. The new technologies as generation sequencing have difficulty to be categorized or used "as is" [1] because of

M. Virvou et al. (Eds.): JCKBSE 2018, SIST 108, pp. 108–116, 2019.
https://doi.org/10.1007/978-3-319-97679-2_11

the vast amounts of data. Over the years there have been a number of applications and methods the scientists can use, in order to validate data and results produced by experiments or projects. More specifically, various techniques, as the combination of biomedical data and computing science have been used to facilitate the scientific research [2]. Bioinformatics is a scientific field whose main purpose is to analyze statistically and categorize the information flow produced by scientific experiments or laboratory work.

Scientists are able to analyze scientific data and evaluate huge data sets by using various methods. These tasks were quite difficult during the past decades On the one hand because of the limited computational power and the lack of an interdisciplinary field which could be used, like Bioinformatics. During the last years vast amount of data, as mentioned above, is starting to accumulate. The main problem is how to utilize these huge amount of data in the future. And for this reason it is imperative to find new techniques. Several attempts have been made, in order to extract information from various scientific papers and/or journals freely available online. We can find various techniques in this way, in the previous years as at Stapley et al. [3]. In this paper the researchers have introduced term "biobibliometrics" to describe the use of bibliometric techniques on papers that are related to biological issues. We can apply the technique in a similar way for educational disorders issues. The specific implementation we can find at Stapley's could verify the bibliometrical connection between biological data as the EEG metrics, based solely on the rate of their common appearances in the abstracts of scientific papers and journals.

In [20] authors combine the learning disability described as "attention deficit hyperactivity disorder (ADHD)" with two EEG metrics, the EEG Rythms and the Event Related Potentials (ERPs). More specifically, in this paper authors associate group differences between ADHD and other populations, the slow EEG rhythms described by Jasper et al. [21] and event-related potentials (ERP).

In this paper, we are going to investigate various methods and finally to implement a custom made IT system in order to find connections between learning difficulties and disabilities and the brain operation and signaling. EEG is the main system for register the brain activity. Thus EEG is a significant medical issue. As first step we implement a search in the free full-text archive of Educational disorders, biomedical and life sciences journal literature at the U.S. in order to investigate the possible association between the selected disorders and the various EEG metrics.

The present paper is divided in the following three parts: The "methodology" in which is explained the methodology we develop. The second part concerns the "results" and the third part includes some discussion and the final conclusion.

2 Experimental and Computational Details

The primary and basic goal of the system will be the ability to search for various EEG metrics supplied by the user and identify any connections or interactions with other metrics based on how frequently they are met together in several papers stored in PubMed central database [4].

2.1 Methodology

Based on the principals of bibliometrics and statistics the system will take into account a series of parameters in order to create a weight-graph between metrics and learning disorders. The basic parameters will be:

- Frequency of the co-appearance of two EEG metrics in the abstracts of papers, freely available online with no restrictions
- EEG metrics Co-citation coupling [15]
- Analysis of related EEG metrics in pairs
- Analysis of the probability of relation between EEG metrics that co-exist in several papers based on the PubMed Central Database.

2.2 High Level Design

The system will constantly poll for EEG metrics and analyze their appearance in papers stored in PubMed [4]. It will then store and link this information when it is requested by the user. For example when metric1 is analyzed the system will store the PID (Paper ID) of PubMed for each paper that contains metric1. Then the same procedure is going to be followed for metric2, metric3 …. metricN. The system based on the user input will construct relations between metrics following the basic principles mentioned above. This routine will be running in real time and will update the information of each gene since the amount of papers being submitted every day could change the final graphs dramatically.

Table 1. Constructing relations between metrics

Metric ID		Metric ID
Metric1		Metric2
PID		PID
0001	→	0001
0003		0002
0005	→	0005
0006	→	0006
0010		0011

As seen on Table 1 we could construct a relation node between Metric1 and Metric2 with the weight of 3.

3 Experimental Results

We are going to use the findings of the previous study [medical hypothesis] to identify possible bibliographic relationship between the various EEG metrics, as EEG Entropy, Power Spectra (PS) and Mental Workload which are associated with the EEG and the metrics EEG Rhythms and ERPs which are associated with the Learning Disorders and the brain [19, 20]. In order to do so, we have applied a searching mechanism via PCM of PubMed services and the results are presented in Tables 2 and 3.

Table 2. Co-appearances between metrics (papers)

	Mental workload	EEG entropy	PS	ERPs	EEG rhythms
Mental Workload 4734 papers		66	230	546	240
EEG Entropy 2650 papers	66		627	960	768
PS 74218 papers	230	627		4919	2713
ERPs 86324 papers	546	960	4919		4466
EEG Rhythms 10382 papers	240	768	2713	4466	

Table 3. Co-appearances between metrics (per cent)

	Mental workload	EEG entropy	PS	ERPs	EEG rhythms
Mental Workload 4734 papers		1,39%	4,86%	11,5%	5,07%
EEG Entropy 2650 papers	2,49%		23,7%	36,2%	29,0%
PS 74218 papers	0,31%	0,84%		6,63%	3,66%
ERPs 86324 papers	0,63%	1,11%	5,70%		5,17%
EEG Rhythms 10382 papers	2,31%	7,40%	26,1%	43,0%	

3.1 Investigating the Relations Between Metrics

The purpose of this subsection is the investigation of possible connections between the PS ERPs and EEG Rythms metrics. For this reason we correlate the possible connection between the mentioned metrics. By analyzing the data from Table 1 and analyzing the connections between the above mentioned metrics we obtained the following results which are depicted in Fig. 1.

According to the above results now it is possible to apply the co-citation normalization procedure [16] which is based on the following equation

$$norm = \frac{|in(PS) \cap in(ERP) \cap in(EEGRythms)|}{in(PS) \cap in(ERP)} = \frac{1474}{4919} = 0.30 \quad (1)$$

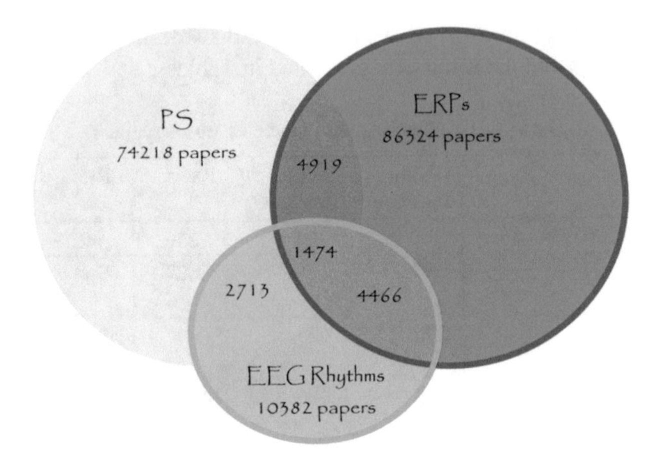

Fig. 1. The intersection between Power Spectra and ERPs with EEG Rhythms

The interpretation of this result indicates that the value 0.30 gives a possible bringing between PS and ERPs metrics at 30% and this lead the ascertainment that a possible research in this issue obtains a higher successful rate than previous attempts.

3.2 Observing the Relation Between EEG Metrics Over Time

The purpose of this subsection is the visualization of the relations we reveal in the previous subsection. Also we can draw conclusions by observing this visualization doing. In order to visualize the relation between metrics Power Spectra, ERPs and EEG Rhythms we are going to investigate the appearance of each metric, in PubMed Central Database, over the past 2 years with one month interval (Fig. 2). We have also determined the relation between EEG Rhythms and Power Spectra, and between Power

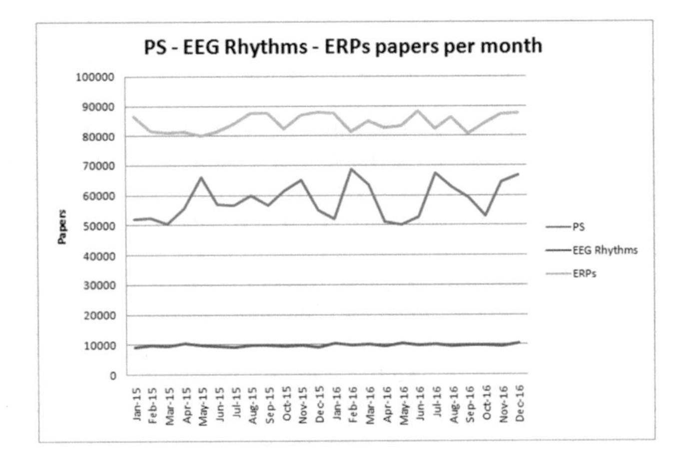

Fig. 2. Appearances of Power Spectra, EEG Rhythms and ERPs in PubMed Central over the past two years.

Spectra and ERPs over the period of 2 years, again with one month interval (Fig. 3). All the data used to construct the graphs in Figs. 2 and 3 can be observed in detail in Table 3.

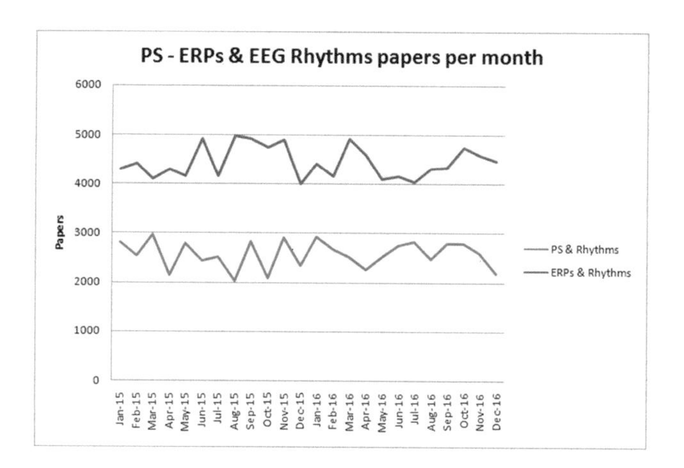

Fig. 3. Number of co-appearances between EEG rhythms and power spectra and between EEG rhythms and ERPs in PubMed central over the past two years.

As we can observe in Figs. 2 and 3 there is a noticeable increase in EEG rhythms appearances in journals, Power Spectra appearance rates seem to be stable and there is a slight increase in the appearances of ERPs. We can also notice that there is a slight increase of the bio-bibliometric relation between EEG Rhythms and Power Spectra. Finally we identify an almost stable relation between Power Spectra and ERPs.

4 Results and Discussion

The bibliometric approach we presented in this paper, between Learning disorders and Human brain via EEG metrics, could provide a very important tool for the scientific community, improving connections between learning disabilities and human brain EEG metrics. In this paper we showed that the EEG Rhythms metric plays a crucial role regarding the identification of possible relation between EEG and learning disorders. Through this paper we showed that if we study the graphs provided we are sure that EEG Rhythms could be the common link between EEG metrics affecting possibility of learning disability existence. Thus we believe that further pursue of this work could be made by taking into account the crucial role of the EEG Rhythms metric.

We have the opinion that the statistical system will become a useful tool for researchers around the world. The specific implementation will provide a mean to connect, seemingly unconnected EEG metrics.

Table 4. Appearance and co-appearances of Power Spectra (PS), EEG Rhythms and ERPs

	PS	EEG Rhythms	ERPs	PS & Rhythms	ERPs & Rhythms
Jan-15	57998	10969	89243	2498	4070
Feb-15	65732	10949	89273	2647	4164
Mar-15	55692	9180	86852	2898	4499
Apr-15	56547	10757	83355	2483	4614
May-15	50204	9426	80970	2704	4672
Jun-15	69401	9685	85062	2084	4358
Jul-15	56685	9163	83456	2045	4123
Aug-15	50845	9103	81966	2226	4540
Sep-15	55984	9724	87419	2127	4278
Oct-15	60444	10403	84359	2511	4464
Nov-15	50911	9676	88556	2384	4985
Dec-15	65308	9527	83098	2185	4078
Jan-16	64269	10259	80840	2512	4938
Feb-16	61005	10946	85410	2596	4420
Mar-16	57105	10583	89531	2820	4931
Apr-16	68946	10390	82848	2340	4029
May-16	59530	10333	88634	2070	4992
Jun-16	60185	10898	88830	2617	4237
Jul-16	57529	9001	85199	2136	4439
Aug-16	67001	10705	82303	2878	4539
Sep-16	69214	9948	81267	2112	4853
Oct-16	57998	10897	88743	2701	4589
Nov-16	62485	10911	82330	2678	4978
Dec-16	68725	10920	88121	2635	4767

5 Conclusions and Future Work

In this paper we approached with bibliometric terms the scientific fields of Education and Electroencephalography (EEG) metrics. More specifically we presented association and communication between human brain Learning disorders and various EEG metrics. We used EEG Entropies, Power Spectra, Event Related Potentials, EEG Rhythms Oscillations. As a conclusion we can conclude with that, via this paper we provide with a very important tool the scientific community, improving connections between learning disabilities and human brain EEG metrics. In this paper we showed that the EEG Rhythms metric plays a crucial role regarding the identification of possible relation between EEG and learning disorders. Based on the graphs provided we are fairly sure that EEG Rhythms could be the guide and the common link between EEG metrics affecting possibility of learning disability existence. For this reason we believe that further pursue of this work could be made by taking into account the crucial role of the EEG Rhythms metric. We have the opinion that the statistical system will become a useful tool for researchers around the world. The specific implementation will provide a mean to connect, seemingly unconnected EEG metrics.

As future work, we plan to develop a parametric information system in order to automate the statistical procedure of Tables 2, 3 and 4. The information system will constitute from an intelligent calculate mechanism based on Python and a graphical user interface with parameters that can automatically extract bibliometric information and statistical information from PubMed or other online databases.

References

1. Metzker, M.L.: Sequencing technologies—the next generation. Nat. Rev. Genet. **11**, 31–46 (2010)
2. Cohen, J.: Bioinformatics—an introduction for computer scientists. ACM Comput. Surv. **36**(2), 122–158 (2004)
3. Stapley, B.J., et al.: Biobibliometrics: information retrieval and visualization from co-occurrences of genenames in Medline abstracts. In: Pacific Symposium on Biocomputing, vol. 5, pp. 526–537 (2000)
4. Martzoukos, Y., Papavlasopoulos, S., Poulos, M., Syrrou, M.: Biobibliometrics & gene connections. In: IISA 2015 (2015)
5. http://www.ncbi.nlm.nih.gov/pmc/. Accessed 04 Oct 2015
6. Poulos, M., Stefanidis, V.: Synchronization of small set data on stable period. In: 2nd International Conference Mathematics and Computers in Science and Industry, Malta (2015)
7. Baars, B.J., Gage, N.M.: Cognition, Brain, and Consciousness: Introduction to Cognitive Neuroscience, 2nd edn. Elsevier, Oxford (2010)
8. Hodgkinson, C.A., et al.: Genome-wide association identifies candidate metrics that influence the human electroencephalogram. Proc. Natl. Acad. Sci. **107**(19), 8695–8700 (2010)
9. Dumermuth, G., Molinary, L.: Spectral analysis of EEG background activity. In: Gevins, A.S. (ed.) Handbook Methods of Analysis of Brain Electrical and Magnetic Signals. Elsevier, Amsterdam (1987)
10. Sommer, B.J., Barycki, J.J., Simpson, M.A.: Characterization of human UDP-glucose dehydrogenase. J. Biol. Chem. **279**, 23590 (2004)
11. Marcel, S., del R. Millán, J.: Person authentication using brainwaves (EEG) and maximum a posteriori model adaptation. IEEE Trans. Patt. Anal. Mach. Intell. **29**, 743–752 (2007)
12. Dougherty, M.K., Morrison, D.K.: Unlocking the code of 14-3-3. J. Cell Sci. **117**, 1875–1884 (2004)
13. Hirsch, L.J., Richard, P., Brenner, R.P.: Atlas of EEG in Critical Care. Wiley-Blackwell, Chichester (2010)
14. Percival, D.B., Walden, A.T.: Spectral analysis for physical applications multitaper and conventional univariate techniques. Cambridge University Press, Cambridge (1993)
15. Small, H.: Co-citation in the scientific literature: a new measure of the relationship between two documents, **24**(4), 265–269 (1973)
16. Li, F., et al.: Applying association rule analysis in bibliometric analysis-a case study in data mining. In: Proceedings of the Second Symposium International Computer Science and Computational Technology (ISCSCT 2009), Huangshan, P. R. China, 26–28, December 2009, pp. 431–434 (2009)
17. Matthews, G., Reinerman-Jones, L., Abich IV, J., Kustubayeva, A.: Metrics for individual differences in EEG response to cognitive workload: optimizing performance prediction. Personality Individ. Differ. **118**(1), 22–28 (2017)

18. Oliveira, A.S., Schlink, B.R., David Hairston, W., König, P., Ferris, D.P.: Proposing metrics for benchmarking novel EEG technologies towards real-world measurements. frontiers in Human Neuroscience
19. Stefanidis, V., Anogianakis, G., Evangelou, A., Poulos, M.: Learning difficulties prediction using multichannel brain evoked potential data. In: IEEE Proceedings, MCSI, pp. 268–272 (2016)
20. Lenartowicz, A., Loo, S.K.: Use of EEG to diagnose ADHD. Curr. Psychiatry Rep. **16**(11), 498 (2014)
21. Jasper, H.H., Solomon, P., Bradley, C.: Electroencephalographic analyses of behavior problem children. Am. J. Psychiatry **95**(3), 641–658 (1938)

Effectiveness of Automated Grading Tool Utilizing Similarity for Conceptual Modeling

Yuta Ichinohe[1(✉)], Hiroaki Hashiura[1(✉)], Takafumi Tanaka[2], Atsuo Hazeyama[3], and Hiroshi Takase[1]

[1] Nippon Institute of Technology, Saitama, Japan
2178003@estu.nit.ac.jp, {hashiura,takase}@nit.ac.jp
[2] Tokyo University of Agriculture and Technology, Tokyo, Japan
s163071w@st.go.tuat.ac.jp
[3] Tokyo Gakugei University, Tokyo, Japan
hazeyama@u-gakugei.ac.jp

Abstract. In recent years, a system engineer is required for advanced modeling because systems have become more complicated. A lecture of model diagrams has been carried out in many universities for training system engineers. The lecture has the problem that teachers cannot give enough feedback to students. The problem is due to a high degree of freedom of the model diagram. The model diagrams have words that have equal meaning and have different expressions. Teachers should consider these features when grading deliverables of students. In addition, a common way to grade deliverables is that teachers check "model diagrams" by hands. So, teachers need a lot of time when grading. Students cannot review their own deliverables. In this research, we develop a tool that gives feedback for students in lectures. We focus on conceptual modeling using notation based on class diagrams. Also, the tool automatically giving feedback for students at near real time. Teachers judge words that have equal meaning and have different expressions. The tool uses these judgements when grading deliverables of students. The content of feedback is a result of grading all operations. The results are visualized by a line chart. We experimented and verified the adequacy of the proposed method. We confirmed the usefulness of the method and the correlation between result of grading and time of unused KIfU.

Keywords: Conceptual modeling · Auto feedback · Visualization

1 Introduction

A system engineer uses a model diagram when modeling a system. The model diagram supports their communication. It has high degree of freedom because natural language is used inside the model diagram. If system engineers use model diagrams, they should learn about its notation. Therefore, a lecture of model diagrams has been carried out in many universities for training system engineers. The lecture has the problem that teachers cannot give enough feedback to students. The model diagrams have words that have equal meaning and have different expressions. Teachers should consider these features when grading deliverables of students. In addition, a common way to grade

M. Virvou et al. (Eds.): JCKBSE 2018, SIST 108, pp. 117–126, 2019.
https://doi.org/10.1007/978-3-319-97679-2_12

deliverables is that teachers check it by hand. Therefore, teachers need a lot of time when grading. On the other hand, novices (defined as "he has never had training designing model diagrams") need feedback from the teachers in real time. Modeling requires "the ability to create a model", "the ability to understand requirements of software" and "the ability to preserve the status when there are changes in the model" [1]. The novices do not have "the ability to understand requirements of software". Therefore, students cannot review their own deliverables.

In this paper, the authors developed a tool that implements the following proposed method. The tool automatically giving feedback for students at near real time. Teachers judge words that have equal meaning and have different expressions. The tool uses these judgements when grading deliverables of students. We experimented and verified the adequacy of the proposed method. In addition, we investigated the correlation between results of grades and the time of unused KIfU.

2 Related Work

Masumoto et al. [2] developed a function that detects the misses in basic conceptual modeling. The function evaluates deliverables of the novices and gives the evaluation results in sentence form. Miyajima et al. [3] developed a tool based on Masumoto's function. Teachers create the correct answer in tabular form before grading, and this tool uses the rule when grading deliverables of students. This tool supports to score items that are difficult detection automatically. This tool gives feedback in sentence form. In this research, feedback is visualized with a line chart based on the similarity between deliverables of student and the correct answer.

Stikkolorum et al. [4] proposed four strategies that novices use to design class diagrams. They confirmed the difference of models between "Depth First Strategy" and "Breadth First Strategy" based on layout and richness. They wanted to confirm if a strategy was excellent or not, and they think meaning of time that students "do nothing". They think the 60 to 200 s meanings are trial and error and the 1000 s meanings are difficult to interpret. In this research, we can confirm if a strategy is excellent by converting because we convert deliverables from elements into F-measures. In addition, we confirmed difference of deliverables depending on the time of trial and error. Therefore, we think the time for trial and error has meaning.

3 Our Approach

We propose a method that can automatically grade deliverables of novices. This method can grade at near real time. The teacher workload is to check words that have equal meaning and have different expressions.

3.1 F-Measure (Similarity)

We use F-measure that is a harmonic mean of precision and recall. It shows the similarity between a student's answer and the correct answer. The definition of precision and recall are as follows:

(1) Precision shows the percentage of correct answer.

$$Precision = \frac{Number\ of\ correct\ novice\ answers}{Number\ of\ novice\ answers} \quad (1)$$

(2) Recall shows the percentage of covers.

$$Recall = \frac{Number\ of\ correct\ of\ novice\ answers}{Number\ of\ possible\ correct\ answers} \quad (2)$$

We selected the following operations for grading:

- CREATE – creation of class.
- UPDATE – update of class name and attribute.
- ADD – addition of attribute and association.
- DELETE – deletion of class, class name, attribute and association.

The maximum value of F-measure is 1.00. If novices' F-measure is near 1.00, they are judged to have high similarity of their own deliverables.

The proposed method uses only class names when grading associations. The F-measures of association are based on class names at both ends of an association. It is difficult for the novices to design association names and multiplicity. Therefore, the method does not consider their association names and multiplicity.

3.2 Grading Tool

Figure 1 shows an overview of the proposed method. The KIfU is a modeling environment that was developed by Tanaka et al. [5]. The tool can log operations (e.g. creation or deletion of a class, update of a class name, movement of a class, etc.) of the students. Also, they can view and reproduce the processes of the operations. We developed the tool uses these logs when grading deliverables of students. Students can draw a conceptual modeling freely. Therefore, teachers should check similar word expressions for grading deliverables. The tool elicits words that have equal meaning and different expressions. If teachers decide to add answers, they register the words in the tool. This work is only registered once for the same word in model. The tool calculates the F-measure after these works are added. F-measures are class name, attribute and association.

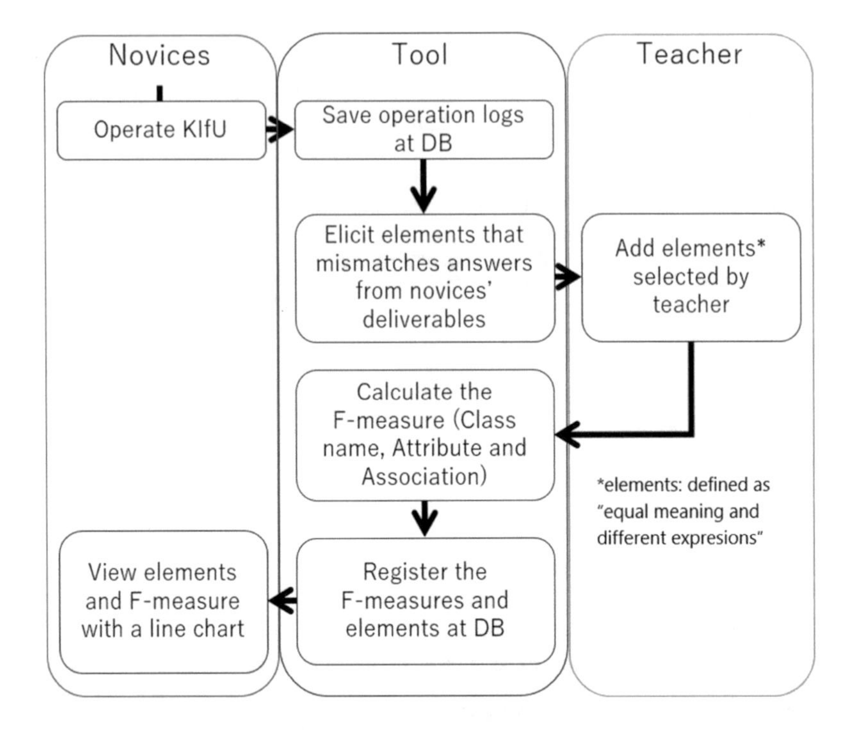

Fig. 1. An overview of the proposed method

Figure 2 shows an example of a line chart based on result. The vertical axis is the F-measure and the horizontal axis is the number of operations. We know that it was completed in 58 operations and all F-measures are 1.00. In this way, novices fix their own models with use of the F-measure.

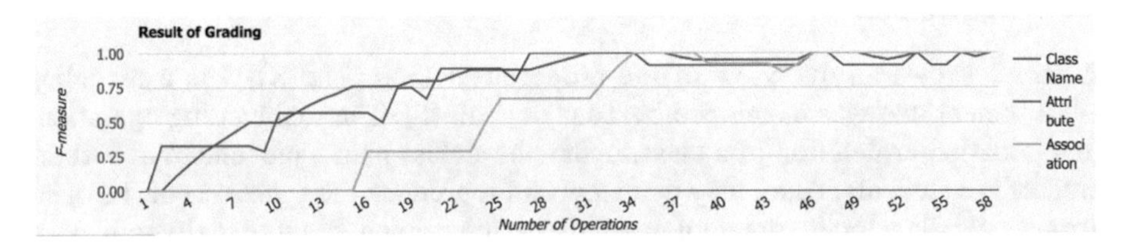

Fig. 2. Visualization of the F-measure

3.3 View Elements

If we input values into the textbox, we can view the elements of the class model at a certain point. Figure 3 shows an example for viewing these elements. The elements are lists of 116th operation. We know that the novice created four classes and three associations.

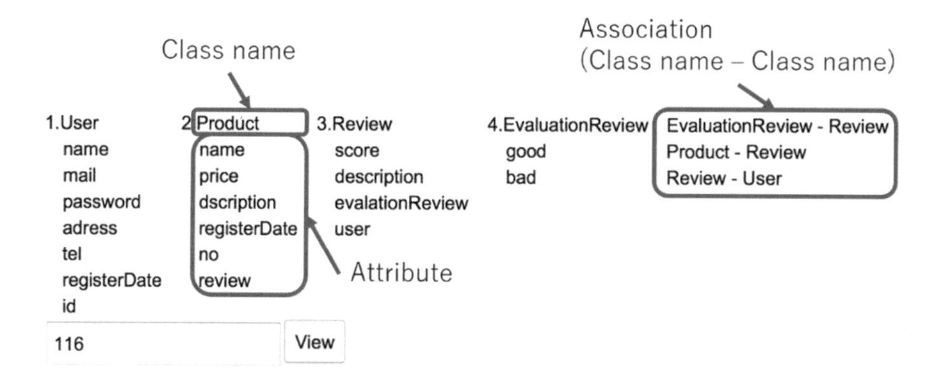

Fig. 3. An example of the class model viewer for student

4 Experiment

We verified the validity of F-measure by conducting experiments with the help of seven students that are majoring in Computer and Information Engineering at Nippon Institute of Technology.

4.1 Research Questions

In advance of our experiment we listed the following two research questions:

- RQ1: Is F-measure useful as a quantitative measure for grading deliverables of students?

We assumed the F-measure is available as tips that help students fix their models. If students use the measure, they can design models similar the correct answer. Also, we think the measure does not need to include the importance of classes.

- RQ2: Is there any correlation between F-measures and results of the experiment?

We assumed the result of the experiment will have certain features. We defined two hypothesis:

- The result have correlation F-measures and numbers of KIfU's operation
- The result have correlation F-measures and numbers of KIfU unused time

We can categorize subjects based on the features. Also, the categories have common F-measure features.

4.2 Method

We used conceptual modeling based on notation of UML (Unified Modeling Language) class diagram. Also, we used KIfU as a design environment. We lectured for 20 min about conceptual modeling because the subjects do not have enough knowledge about it.

First, subjects created a model with only KIfU for 40 min (without the tool). Next, they fixed their own model with feedback (F-measures and elements of class model) from the tool for 40 min (using the tool). We prohibited the subjects from consulting with each other. Prevention of mixing with a third party is the purpose of this rule.

4.3 Result of Experiment

A paired samples t-test was conducted to compare "without the tool" and "using the tool". The significance level α used in t-test is 0.05.

Figure 4 shows changes of the F-measure about "without the tool" and "using the tool" in class name. There was a significant difference in the F-measures of class name; $t(6) = -0.788$, $p = 0.001$.

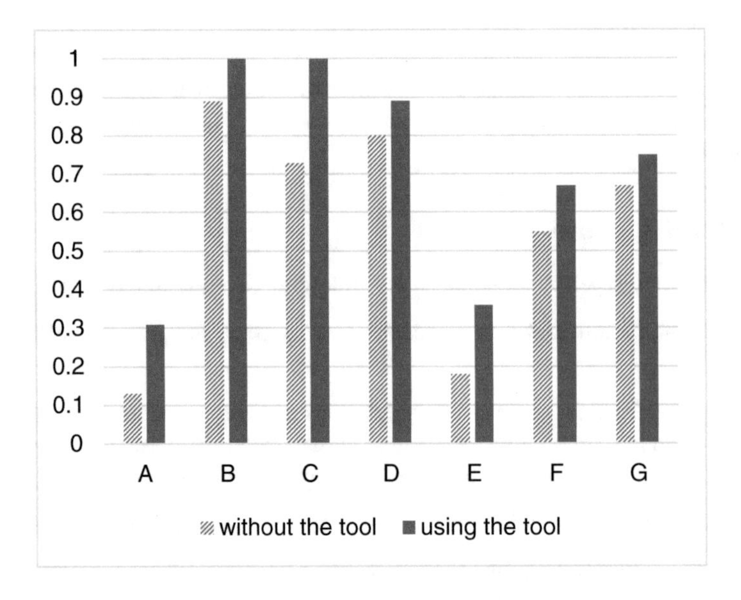

Fig. 4. Comparison class names of F-measure

Figure 5 shows changes of F-measure with "without the tool" and "using the tool" in association. There was a significant difference in the F-measures of association; $t(6) = 2.951$, $p = 0.025$.

Fig. 5. Comparison associations of F-measure

Figure 6 shows the number of operations with "without the tool" and "using the tool". There was not a significant difference in the KIfU's number of operations; $t(6) = 1.681$, $p = 0.156$. Also, Fig. 7 shows unused time of KIfU. Unused time of subject A is the longest in all subjects' one unused time. Subject E's number of unused time is the most in all subjects. Figure 8 shows the numbers that are operations of unused time about KIfU. We set over 61 s because 60 to 200 s meaning are trial and error [4].

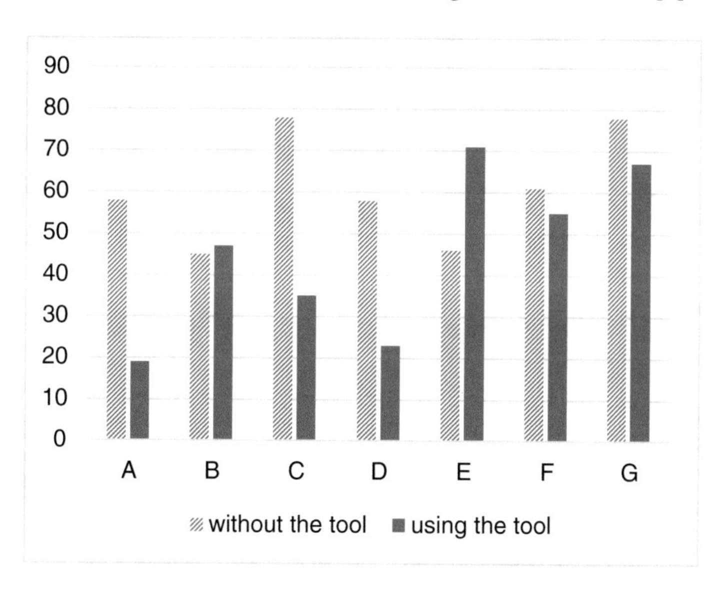

Fig. 6. Comparison of KIfU's number of operations

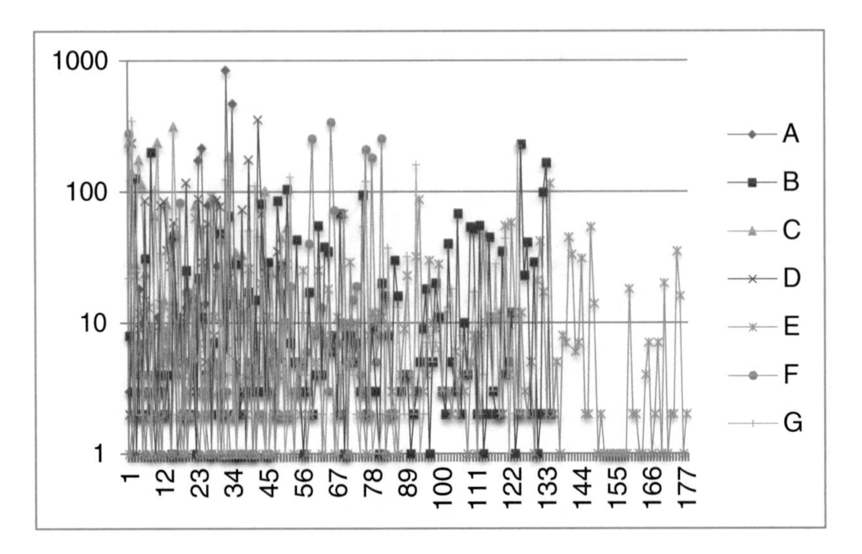

Fig. 7. Unused time of KIfU

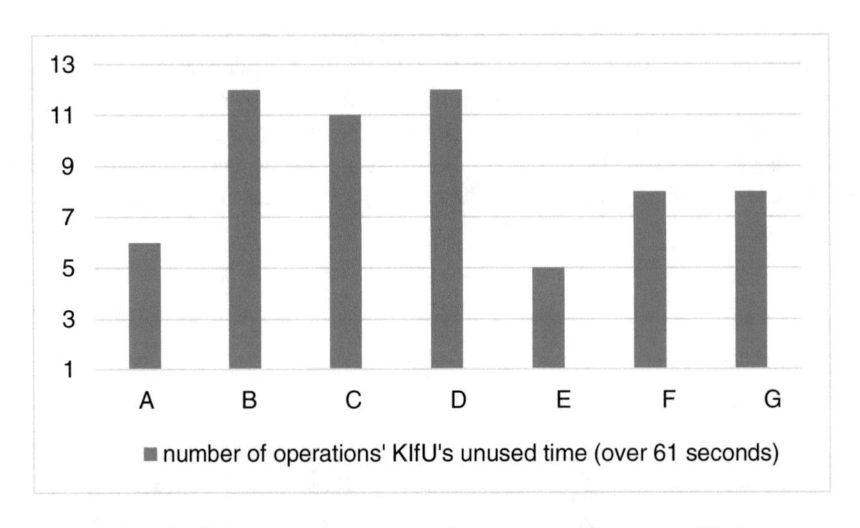

Fig. 8. The number that are operations of unused time about KIfU

5 Discussion

5.1 RQ1: Is F-Measure Useful as Quantitative Measure for Grading Deliverables of Students?

In Fig. 4, F-measure of "using the tool" is higher than "without the tool". Also, in Fig. 5, F-measure of "using the tool" is higher than "without the tool" except for subjects A and E. We think these results show subjects could fix the models with the use of feedback. Therefore, the F-measure is useful when students fix models. On the other hand, we think importance of classes have margins. So, if we decide grades based on F-measure, we should include it. We want to confirm the difference between F-measure (based on precision and recall) and F-measure (based on precision, recall and importance of each classes).

F-measure of subjects A and E are lower than other subjects. The problems are due to set name of operation in class name (e.g. correct class name is “user”, but subjects set “user register” in it). Also, the proposed method needs two classes when grading associations because the F-measure of association are based on class names at both ends of an association. Thus, F-measures of subjects A and E are low value because their numbers of correct class names are less than the other subjects (Fig. 5). If we want to support them, we need to add other feedback in addition to F-measure. We want to consider other feedback (e.g. other measures, other functions, and so on).

5.2 RQ2: Is There Any Correlation Between F-Measures and Result of the Experiment?

In Fig. 6, subject A’s number of operations are the least of all subjects (in “using the tool”). We think he took a lot of time with trial and error. Also, subject E’s number of operations are the largest of all subjects (in “using the tool”). We think he used a lot of time to move his hands. But, we could not confirm a correlation between their features and F-measure (Figs. 4 and 5). On the other hand, the F-measure of subjects B and C are larger than other subjects (Figs. 4 and 5). But, we could not confirm a correlation between their F-measures and KIfU’s number of operations (Fig. 6). Therefore, we think the results of this experiment do not have correlation of the F-measure and KIfU’s number of operation.

In Figs. 4 and 5, the F-measure of subjects B, C and E are larger than other subjects. In Fig. 8, their unused time of the KIfU (over 61 s) are larger than the other subjects. We think they used a lot of time to trial and error (e.g. viewing their own process of editing, thinking of things to do next, and so on). Next, the values of subjects F and G are large values (Fig. 8). Their F-measures are large after subjects B, C and D (Figs. 4 and 5). Also, the values of subjects A and G are lower than other subjects (Fig. 8). Their F-measures are lower than other subjects (Figs. 4 and 5). Thus, we can categorize the subjects based on these results. We defined features that if the F-measure of a student was larger than other students, his time of unused KIfU’s are larger than the other students. We think subjects used a lot of time to read the problem and/or using the tool developed by the authors. But we didn’t develop a function logging the operations of the tool. In this future, we want to develop this function, and investigate the work of unused time of the KIfU.

In addition, we want to define the time of unused KIfU. If we selected over 51 s (Fig. 7), the numbers (Fig. 8) of subject E is larger than other, except subject B. Also in Fig. 7, the unused time of subject A is the longest in all subjects (about over 800 s). But, we could not confirm the correlation between the feature and F-measures (Figs. 4 and 5). We think the problem is due to set time restrictions (40 min) in this experiment. We want to verify the adequate time of unused KIfU. Additionally, we want to conduct an experiment without time restrictions.

6 Summary

6.1 Conclusion

In this paper, the authors developed a tool that can give feedback at near real time.

We experimented and confirmed the adequacy of the tool and the method. The method is useful when students fix their own models. We defined hypothesis that have correlation between F-measures and numbers of KIfU unused time.

6.2 Future Work

In the future, we want to confirm the adequacy of the hypothesis (e.g. increase number of subject, increase problems, and so on). In addition, we want to confirm any correlation between exams and F-measure in the actual lecture. Also, we want to investigate the work of students' unused time the KIfU.

Acknowledgments. This work was supported by JSPS KAKENHI Grant Number JP17K00475, JP18K11579. We would like to thank the members of Hashiura's Laboratory at Nippon Institute of Technology.

References

1. Sendall, S., Kozaczynski, W.: Model transformation: the heart and soul of model-driven software development. In: IEEE Software, vol. 20, pp. 42–45. IEEE (2003)
2. Masumoto, K., Kayama, M., Ogata, S., Hashimoto, M.: Developing an error detecting tool based on quantitative error analyses for novices learning to create error-free simple class diagrams. In: The Special Interest Group Technical Reports of Information Processing Society of Japan, IPSJ, vol. 2015-SE-187, No. 15, pp. 1–7 (2015). (In Japanese)
3. Miyajima, K., Ogata, S., Kayama, M., Okano, K.: A support tool for rule-based scoring of class diagrams in UML modeling education. In: The Institute of Electronics, Information and Communication Engineers Technical Report, vol. 115, No. 154, pp. 149–154 (2015). (In Japanese)
4. Stikkolorum, D.R., Ho-Quang, T., Chaudron, M.R.V.: Revealing students' UML class diagram modelling strategies with WebUML and LogViz. In: 2015 41st Euromicro Conference on Software Engineering and Advanced Applications (SEAA), pp. 275–279. IEEE (2015)
5. Tanaka, T., Hashiura, H., Hazeyama, A., Komiya, S.: Do learners to create an artifact with good quality make a number of trials and errors during the editing process? In: 2015 3rd International Conference on Applied Computing and Information Technology/2nd International Conference on Computational Science and Intelligence (ACIT-CSI), pp. 28–35. IEEE (2015)

Generating Scenarios with Access Permission from a Conceptual Model

Takako Nakatani[1(✉)], Hideo Goto[2], Osamu Shigo[3], and Taichi Nakamura[2]

[1] The Open University of Japan, Chiba, Japan
tinakatani@ouj.ac.jp
[2] National Institute of Informatics, Tokyo, Japan
[3] Tokyo Denki University, Tokyo, Japan

Abstract. There are methods to cope with the incompleteness of requirements. This paper proposes a conceptual model based method to generate scenarios. The method assumes that the unfriendliness of systems comes from the gap between the operational model and the mental model of users. The operational model can be represented in use case models or users' scenarios. Our method applies a conceptual model with class diagrams as a part of the mental model of users. We also develop a tool as a bridge between these models. It automatically generates users' scenarios from the conceptual model with access permissions. In order to evaluate the effectiveness of the method, we study a case and clarify whether the method and the tool contribute to the improvement of the completeness of requirements. The case focused on is a system of a domain of academic affairs of the Open University of Japan. As the result, the tool implemented could generate new scenarios that have not been implemented within the system.

Keywords: Requirements engineering · User scenario
Conceptual model

1 Introduction

There are various problems in eliciting and writing requirements specifications. Engineers can apply methods to elicit requirements [3, 5, 13, 17, 18]. Though those techniques contribute to the improvement of software quality, requesters still now claim that the delivered system does not satisfy their demand. Such a situation is inevitable, since it is hard to elicit requirements completely. In this paper, we introduce a method and a tool to cope with the completeness of requirements.

The following conventional solutions are partly useful to solve the problems of incompleteness. First, requirements analysts acquire complete domain knowledge. Second, domain ontology helps analysts find unknown requirements. Third, persona analysis is effective in the analyzation of a typical users' behavior. Further, analysts define requirements with scenario analysis and use case modeling; nevertheless, some requirements are left to be elicited.

M. Virvou et al. (Eds.): JCKBSE 2018, SIST 108, pp. 127–136, 2019.
https://doi.org/10.1007/978-3-319-97679-2_13

In order to develop a method to improve the completeness of requirements, we focus on web-based systems, as web-based systems have become popular. It is common that users utilize the systems without reading manuals. However, not a small number of systems provide unfriendly functions. We regard that, the "unfriendly" comes from a mismatch between the operational model of the system and the mental model of users. If we resolve the mismatch, we will be able to improve the completeness of requirements. In conventional software engineering methods, we write operational models with use cases, or scenarios. If we develop a mental model and derive the user's scenarios, we will be able to compare them with operational models and evaluate the adequacy of the operational models.

There is however one other problem: when we develop a web system, we cannot focus on requirements of a certain person or a typical person, in other words, persona, because, the users of a web system are diverse. Therefore, the mental model must belong to a general user, but not a particular user. We define general users as those who understand the target domain correctly and traverse the system widely based on their mental model. The mental model can be represented as the structure of the domain in which the users traverse in their mind and physically traverse among the web pages. Therefore, the structural model of the domain is useful to represent the mental model of general users. We call this structural model a *conceptual model*.

The purpose of this paper is to improve the completeness of requirements. We propose a method to elicit requirements for traversing multiple web pages inside the system. These requirements derive from the conceptual model of a target domain. We will show the effectiveness of the method by comparing the elicited requirements and the requirements within actual systems. The study was conducted on a case named *WAKABA*, which is a system of academic affairs of the Open University of Japan.

The structure of this paper is as follows. In Sect. 2, we introduce related work for eliciting requirements. In Sect. 3, we will present an overview of the method and, the tool of the method. In Sect. 4, we introduce a case and present the conceptual model of the domain of the case. We also describe the results of the case study. After Sect. 5, we discuss the effectiveness of the method. In the last section, we conclude this paper.

2 Related Work

When it is anticipated that various users will be indirectly participating in the developing system, a requirements analyst cannot conduct interviews with those users. In such a case, the analyst is recommended to define a typical user as a persona and analyze the behavior of that persona [1]. This method is called, "Persona analysis." Persona analysis is an effective method to elicit interactions between a person and the developing system [4]. However, it is inadequate in its reach for the eliciting requirements of systems so that the requirements satisfy the goals of various kinds of users. Especially for web systems, requirements analysts should take into account various people as stakeholders. We need additional methods for eliciting requirements.

In requirements engineering, the activities of a persona are defined as scenarios. Scenarios are written in natural language, so that they contribute to the improvement of the readability of requirements specifications. On the other hand, scenario analysis has a drawback, that being, it has less evaluation processes with regard to the completeness of scenarios. Use case modeling [8] has a similar problem, in that, we cannot determine whether the use case model covers all of the users' requirements or not. Conceptual modeling has also been utilized for requirements elicitation. Rolland pointed out the importance of conceptual models for the improvement of the quality of requirements specifications [14]. We define a conceptual model as the source of requirements, so that we evaluate the completeness of requirements.

Applying ontology is an important approach for requirements elicitation. According to Studer's definition, "An ontology is a formal, explicit specification of a shared conceptualization" [16]. Ontology has three layers; i.e. domain, inference, and task. Kaiya et al. utilize the domain and inference layers in their requirements elicitation method named ORE (Ontology based Requirements Engineering) [9]. ORE is able to help analysts elicit functional requirements. Our focus is to construct user scenarios with a conceptual model.

Further, Nalchigar et al. introduced a conceptual modeling framework by focusing on information systems that mainly manipulate data [12]. The aim of the framework is to bridge a gap between business strategy and developing systems. Insfrán also applied conceptual models for the prediction and elicitation of requirements [7]. The final product of their method is a program source code. The method may improve the productivity of development, but it does not have a mechanism to elicit coarse-grained requirements: e.g. requirements with multiple use cases.

Nakatani was attempting to predicate requirements changes and explore undefined requirements with a conceptual modeling technique [11]. Goto et al. developed an event list generation system by extracting scenarios from class diagrams [6]. However, it being the case that multiplicities and access permissions are omitted, we extended their logic with regard to the omitted issues.

Users of web systems tend to traverse among numerous pages. Thus, we need to analyze their activities and integrate use cases with relations among the objects that are manipulated in each page. In order to analyze the users' activities, we build a mechanism to derive object-links from a conceptual model. In this paper, we introduce a method to define access scenarios based on connections among objects within a conceptual model.

There are methods to construct conceptual models, such as that introduced by Montes, which is a method that creates conceptual models from use cases automatically [10]. Their method is a traditional object-oriented method: firstly, find objects in use cases and secondly, define relations based on use case scenarios. Augusto proposed a method that translates business process model to conceptual model, rather than create a structural model (conceptual model) directly [2]. Creating and applying conceptual models is a useful means for the understanding of the target world and defining requirements. In this paper, we discuss methods to create conceptual models less, and focus on a way to create users' scenarios from a conceptual model, and as a result, improve the completeness of requirements.

3 A Method Overview

3.1 Approach

In conceptual models, there are classes and relationships between classes which are the abstraction of objects in the world. We apply class diagrams in UML to represent conceptual models.

We can traverse from one class to another class via multiple classes. A traverse pass represents a user's possible exploration among the web pages within the developing system. If a user is permitted to access an object, they can reach the web page of the object. We take into account the permission of each user's group to access the web page of each object. There are four types of permission; create, read, update, and delete. However, "update" and "delete" cannot be executed to access the target object. If a user creates a new object, the user has to be able to access another object, which in tern, has a relationship with the new object. Therefore, we focus on "read" permission, only.

If a users' group is allowed to access one object from other object, a scenario of a user of the users' group can be generated for the access. Hence, we can build a traverse path as a scenario for the users' group from one class to the termination class via multiple classes. The "termination class" means that the class does not have any relationship to other classes, or, the class only has a relationship with other classes that have been visited in the traverse pass.

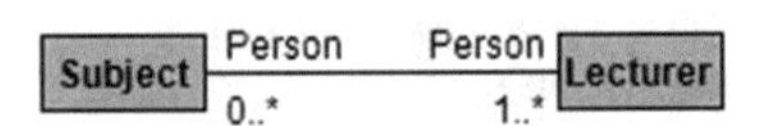

Fig. 1. An example of scenarios integrated functional requirements.

Here is an example of functional requirements: *A person can know the lecturers of a subject*, as shown in Fig. 1. Analysts usually define the accessing method in *Subject* class in a UML's diagram. In object-oriented way of thinking, however, we do not clearly define the permission. We approach the role of class diagrams from a different angle. If an association relationship is defined between *Subject* and *Lecturer* with 0..* to 1..* multiplicity, and every person is permitted to access the association, then every person must be allowed to traverse from an instance of *Subject* to the lectures of the subject; thus, the people can traverse back from one of the lecturers to the lecturer's other subjects. In this way, we can generate scenarios from class diagrams with information of the multiplicity and the permissions for each users' group.

We can define use case description as, "After the user inputs the name or the ID of a subject, the system shows lecturers of the subject." The authorization of every user may be defined as one of the non-functional requirements in the use case. Our question then is; "What kind of operations are allowed to the user who gets lecturers?" We will answer these questions with conceptual models. Further, in order to answer these questions, we need to start our next traverse from *Subject* class. A lecturer must have research results A user who wants to register for one of the subjects of the lecturer is required to access the articles via the lecturer object. In this way, we can construct scenarios by tracing the association relationship or aggregation relationship in class diagrams. Such traces will be terminated when the traverse path visits the same class again.

Though standard use cases can represent functional requirements, sometimes the meanings of the access paths or processes of users' operations are ignored. We regard this situation to be the cause of missing requirements. Our scenarios will be able to include all of the access paths of users, if and only if the conceptual model is adequate. We do not discuss the adequacy of conceptual models in this paper.

The process of our method is as follows:

1. Develop a conceptual model agreed on by domain experts.
2. Define users' groups in a hierarchical structure with inheritance.
3. Define authorizations to access the relationships for every users' group.
4. Extract accessibility of classes on the basis of *associations* and *aggregations* between classes in the conceptual model.
5. Traverse classes by traversing associations and aggregations via classes for each permitted user.
6. For each users' group, construct a scenario with the traverse paths.

Here are three rules for developing conceptual models with UML.

- An association is used to permit access associated objects by bi-directions.
- An aggregation is used to permit access aggregated objects according to the direction of the aggregation.
- Roles defined on edges of an association or aggregation are used to represent permitted users' group.

3.2 ScenaGCon: Scenario Generation Tool from Conceptual Model

In order to generate scenarios from class diagrams, we extended an editor of class diagrams and developed **ScenaGCon** to generate scenarios from conceptual models. **ScenaGCon** outputs scenarios tracing classes focusing on permitted users. The tool image is shown in Fig. 2.

Figure 3 represents an example with *association*, *aggregation* and *association class*. We can generate accessing scenarios from the conceptual model in Fig. 3 as follows:

- Actor M accesses a/an E object, the A objects from the E object, the D object of the A object, the B objects from each A object, and the A object from each B object.

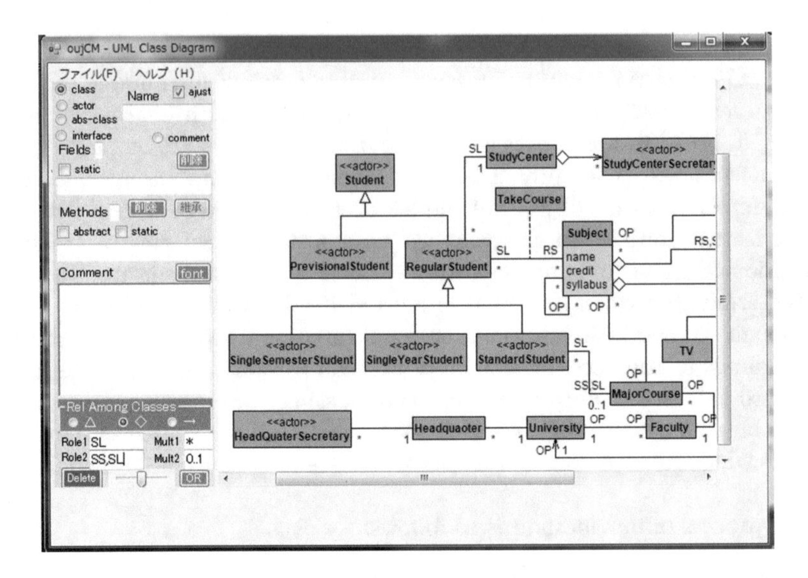

Fig. 2. An image of the tool.

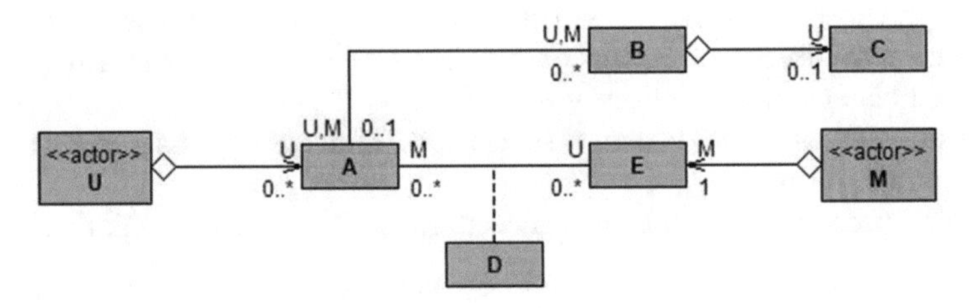

Fig. 3. An example of conceptual models with *association*, *aggregation* and *association class*

- Actor U accesses A objects, the B objects from each A object, and the A object from each B object.
- Actor U accesses A objects, the B objects from each A object, and the C object from each B object.
- Actor U accesses A objects, the E objects from each A object, and the D object of each E object.

The permission of accessing an association is presented at the edge of the association and is paired with the multiplicity.

4 Case Study

4.1 Procedures

We selected a case for the evaluation of our approach. If the existing system lacks of requirements that our method can elicit, we can conclude that our method is effective in eliciting undefined requirements.

The selected case was *WAKABA*, which is a system of academic affairs of the Open University of Japan (OUJ). We can access the site freely, and furthermore, the domain of academic affairs is well known for us. This means that we can be the domain experts and build the conceptual model of the target domain by ourselves.

After building the conceptual model, we generate scenarios by applying ScenaGCon. After that, we tried to use the web pages of *WAKABA* and created a site map of the system. The map represents the structure of objects within the system. We compared the scenarios derived from the conceptual model by ScenaGCon with the map and evaluated the effectiveness of our method in improving the completeness of requirements.

4.2 The Overview of the Case

WAKABA is an OUJ web system of academic affairs. OUJ provides broadcasting lectures for Japanese people. Students can enroll and register lectures via *WAKABA*. So, we can expect students to traverse web pages provided by OUJ in order to search for information on subjects, lecturers, study centers, etc. The major users of *WAKABA* are over 85 thousand students, 360 lecturers, and staffs. Furthermore, Japanese people older than 15 years old are allowed to be students of OUJ. The students of OUJ are highly diversified. Therefore, it is not adequate to define a typical student as a persona for eliciting requirements.

4.3 Analysis of the Result

Figure 4 represents the conceptual model of the world of OUJ's study supporting system and users. The permission of users for each association is shown by abbreviation of the name of users' group. For example, "RS" is the abbreviation of RegularStudent. Class SingleYearStudent inherits class RegularStudent, users who belong to SingleYearStudent inherit the permission of RegularStudent.

We built a site map of *WAKABA*, as there are not any links between subjects and lecturers. This situation tells us that users of *WAKABA* cannot access any lecturer's information. The web pages seem to be designed to inform OUJ's services rather than give information on education.

For example, all users can get syllabi for all subjects. Furthermore, though they can access the names of lecturers from a syllabus, they cannot access the list of articles of the lecturers, nor subjects of them. In contrast, **ScenaGCon** can output a scenario as follows:

Scenario: *Actor OrdinaryPerson accesses a/an University object, the Faculty objects from the University object, the MajorCourse objects from each Faculty object, the Subject objects from each MajorCourse object, the SubjectLecturer objects from each Subject object, and the article objects from each SubjectLecturer object.*

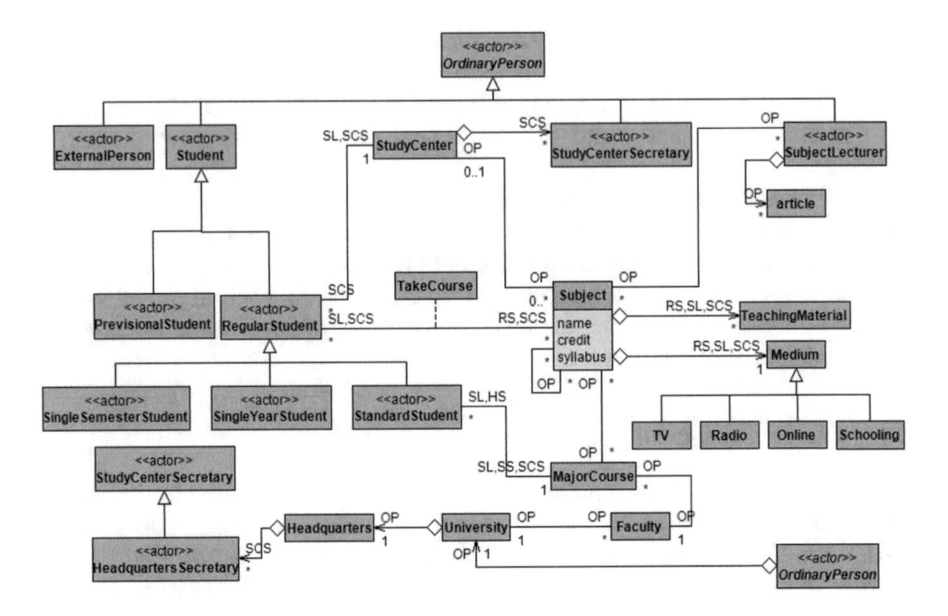

Fig. 4. The conceptual model of OUJ's study supporting the world.

5 Discussion

In this paper, we evaluated the effectiveness of our method and **ScenaGCon** through the case study. We focus on "read" permission of each users' group of objects and links between objects with a conceptual model. As a result, we were successful in elicit functional requirements that were not included within the existing system. However, the outcome is the result of having applied our method to the only one case. In order to conclude that our method and the tool are actually effective to improve requirements completeness, we should apply them other cases. In the rest of this section, we validate the results of the case study [15].

5.1 The Threats of Internal Validity

One threat of internal validity is that the conceptual model of *WAKABA* was built for scenario generation, only. The model is applicable as a general conceptual model of the domain of academic affairs with minor changes. For example, the OUJ specific organizational objects: e.g. various kinds of students, study center, etc., must be deleted and the organizational objects of the target university will be added to the model. Thus, the conceptual model of *WAKABA* is not built only for the scenario generation. If we can get more general conceptual models of other domains, we would analyze the access permission of each users' group and add the information to the model. The key factors with regard to generated scenarios are the relationships between objects and "read permission."

5.2 The Threats of External Validity

The threats of external validity are as follows:

- The undefined requirements that the ScenaGCon could generate were known for the stakeholders of *WAKABA*, but were deleted because of their uselessness.
  We cannot generate a counter-argument to the threat. The strategy of *WAKABA* might be to provide information concerning the structure or process of education, rather than provide information of educational contents efficiently. The scenarios that were generated by ScenaGCon were not able to be found by use case analysis easily. We did not argue pre-condition and post-condition of the scenarios, but accessing user scenes of searching information processes. We made a bridge between use cases as, the operational model and the conceptual model as the mental model of users.
- If the size of the conceptual model becomes too large, the number of scenarios will be exploded.
  That being the case, if the size of the model does become too large, we can divide them into smaller sized models.
- The undefined requirements in *WAKABA* may be able to be found in other systems of universities beyond those of OUJ.
  We researched three universities in Japan and have sought syllabi. The lecturers' homepages could not be reached from the various syllabi acquired. Further, of those acquired, the syllabi were provided in PDF without any hyperlinks. According to the results, we are sure that systems of academic affairs may have similar structure and lack of requirements.

6 Conclusion

In this paper, we propose a method and a tool to construct scenarios by means of a conceptual model. The focus of our research is on the web-system and to follow the traverse process of users among the web pages within the system. According to the case study, we concluded that we could elicit unknown requirements that the actual system did not support. We will apply the method and tool to other domains and evaluate the effectiveness more widely. We will apply the method and tool to another application domain as our future work.

References

1. Aoyama, M.: Persona-and-scenario based requirements engineering for software embedded in digital consumer products. In: The IEEE International Requirements Engineering Conference, pp. 85–94. IEEE Computer Society (2005)
2. Augusto, A., Conforti, R., Dumas, M., La Rosa, M., Bruno, G.: Automated discovery of structured process models: discover structured vs. discover and structure. In: Comyn-Wattiau, I., Tanaka, K., Song, I.Y., Yamamoto, S., Saeki, M. (eds.) Conceptual Modeling, pp. 313–329. Springer, Cham (2016)

3. Chung, L., Nixon, B.A., Yu, E., Mylopoulos, J.: Non-Functional Requirements in Software Engineering. Kluwer Academic Publishers, New York (1999)
4. Cooper, A.: The origin of personas (2003). https://www.cooper.com/journal/2003/08/the_origin_of_personas
5. Dardenne, A., van Lamsweerde, A., Fickas, S.: Goal-directed requirements acquisition. Sci. Comput. Program. **20**, 3–50 (1993)
6. Goto, K., Ogata, S., Shirogane, J., Nakatani, T., Fukazawa, Y.: Support of scenario creation by generating event lists from conceptual models. In: 2015 3rd International Conference on Model-Driven Engineering and Software Development (MODELSWARD), pp. 376–383 (2015)
7. Insfrán, E., Pastor, O., Wieringa, R.: Requirements engineering-based conceptual modeling. Requirements Eng. **7**(2), 61–72 (2002)
8. Jacobson, I., Christerson, M., Jonsson, P., Overgaard, G.: Object-Oriented Software Engineering. Addison-Wesley, Reading (1992)
9. Kaiya, H., Saeki, M.: Using domain ontology as domain knowledge for requirements elicitation. In: 14th IEEE International Requirements Engineering Conference (RE 2006), pp. 189–198 (2006)
10. Montes, A., Pacheco, H., Estrada, H., Pastor, O.: Conceptual model generation from requirements model: a natural language processing approach. In: Kapetanios, E., Sugumaran, V., Spiliopoulou, M. (eds.) Natural Language and Information Systems, pp. 325–326. Springer, Heidelberg (2008)
11. Nakatani, T., Tsumaki, T.: Predicting requirements changes by focusing on the social relations. In: Proceedings of the 10th Asia-Pacific Conference on Conceptual Modeling, pp. 65–70. Australian Computer Society (2014)
12. Nalchigar, S., Yu, E., Ramani, R.: A conceptual modeling framework for business analytics. In: Comyn-Wattiau, I., Tanaka, K., Song, I.Y., Yamamoto, S., Saeki, M. (eds.) Conceptual Modeling, pp. 35–49. Springer, Cham (2016)
13. Robertson, S., Robertson, J.: Mastering the Requirements Process. Addison-Wesley, New York (1999)
14. Rolland, C.: From conceptual modeling to requirements engineering. In: Proceedings of the 25th International Conference on Conceptual Modeling, ER 2006, pp. 5–11. Springer, Heidelberg (2006)
15. Runeson, P., Höst, M.: Guidelines for conducting and reporting case study research in software engineering. Empirical Softw. Eng. **14**(2), 131 (2008)
16. Studer, R., Benjamins, V.R., Fensel, D.: Knowledge engineering: principles and methods. Data Knowl. Eng. **25**(1–2), 161–198 (1998)
17. Yu, E., Giorgini, P., Maiden, N., Mylopoulos, J. (eds.): Social Modeling for Requirements Engineering. MIT Press, Cambridge (2011)
18. Yu, E.S.K.: Towards modelling and reasoning support for early-phase requirements engineering. In: Proceedings of the 3rd International Symposium on Requirements Engineering (RE 1997), pp. 226–235. IEEE (1997)

A Knowledge Transfer Support System from Text-Based Work Reports with Domain Ontologies

Ryutaro Nambu(✉), Kohei Suehiro, and Takahira Yamaguchi

Keio University, 3-14-1 Hiyoshi, Kouhoku-ku, Yokohama, Kanagawa 223-8522, Japan
{taro-taro,kosuehiro}@keio.jp, yamaguti@ae.keio.ac.jp

Abstract. Recently, there is growing interest in knowledge transfer from human experts to novices in order to keep organizational competence. The workers of organizations usually make work reports but they are text-based and not structured. Therefore, they are hard to be reused by other workers.

In this paper, we present how to make domain ontologies to generate structured data form work reports and develop a knowledge transfer support system, combining structured data with domain ontologies. In order to evaluate the system, as real case studies, we take troubleshooting work in an expressway maintenance management company. The evaluation shows us the systems works well with questionnaires to the workers who attend the case studies.

Keywords: Knowledge management · Knowledge transfer · Domain ontology

1 Introduction

Recently knowledge transfer from experts to novices becomes important in organizations. The workers of organizations make work reports in order to keep logs of what they exactly did and share knowledge within the organizations. Therefore, making work reports works for knowledge transfer. However, most work reports are merely accumulated and there are few cases to be reused by other workers. The reason is why work reports shows worker's experiences just in plain (not structured) text way and so are hard to be understood and be reused by other workers [1].

Here in this paper, we take troubleshooting work on the electronic toll collection (ETC) lane as real case studies. We discuss how to structure work reports and develop a knowledge transfer support system and then how the proposed system goes well form the point of facilitating knowledge retrieval with ease.

2 Related Works

In ontology engineering, Razmerita [2] proposed the ontology-based user modeling. In knowledge management, Ido [3] constructed the domain ontology and used it to detect problematic specifications of industrial machines. Athira [4] developed an ontology-based question answering system for domain knowledge.

M. Virvou et al. (Eds.): JCKBSE 2018, SIST 108, pp. 137–146, 2019.
https://doi.org/10.1007/978-3-319-97679-2_14

On the other hand, in this paper we focus on how to convert work documents to structured data and propose the system for acquiring knowledge by giving a structure to the information of work reports.

3 Domain Ontology

3.1 Usefulness of Domain Ontology for Knowledge Transfer

A domain ontology is what specifies and structures the meaning of terms for the important elements of a target domain [5, 6]. The important elements, in other words, domain ontology modeling, depend on what goal a domain ontology works for.

A domain ontology has a class hierarchy and a set of property semantics are defined there. Using class instances and a set of property, working reports should be structured into RDF triple network. Thus RDF triple network enables end-users to access in various ways depending on their concerns; it goes better for knowledge transfer, compared with text-based working reports.

3.2 Definition of Domain Ontology

Considering the goal of troubleshooting support taken as real case studies later, what normal structure, function and behavior a target device has and also how mal-structure, mal-function and mal-behavior go in the case of device in trouble. The former should be specified as a device ontology and the latter as a device in trouble ontology and an alarm ontology in this case.

Device Ontology. Looking at working reports, the workers seem to grasp the devices by their functions. Thus the class hierarchy for a device ontology has been invented, using the device functions. The instances of the device class are concrete devices taken in ETC. See examples in Fig. 1.

Device in Trouble Ontology. When a device goes in trouble, it might have three views: mal-structure, mal-function and mal-behavior. So we specify a device in trouble ontology with three views. Extracting the information related with them from working reports, we invent three kinds of sub-class hierarchy, decomposing abstract concepts to concrete concepts with three views. Then we invent a set of property necessary to describe device in trouble from three views. Trouble cases from working repots should be described with the property set and work much for knowledge transfer from some workers to other workers. See examples in Fig. 1.

Alarm Ontology. Although mal-behavior is the behavior of device in trouble and does not care about making it explicit to humans, an alarm is a kind of information that notifies that devices break down. ETC has explicit mechanism to generate an alarm when the device fails. Alarm is the first step to start with troubleshooting because workers usually identify which device in ETC break down based on the alarm information. Therefore, alarm class hierarchy is related with the device class hierarchy. Like the device ontology,

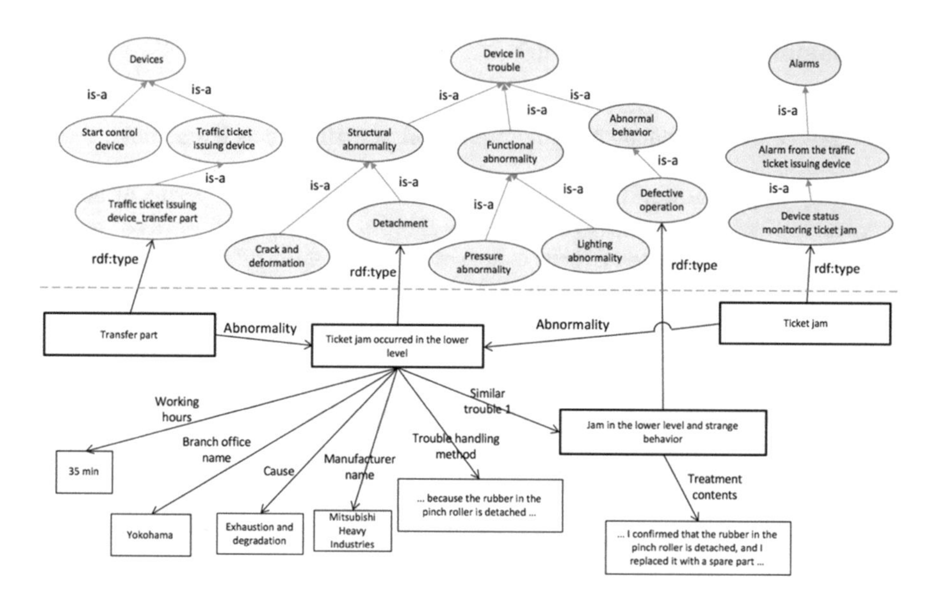

Fig. 1. Device, device in trouble and alarm ontologies.

the instances of the alarm ontology are connected with those of device in trouble ontology. See examples in Fig. 1.

3.3 Structuring Working Reports with RDF Triples Using Domain Ontologies

As shown in Fig. 1, the trouble "Ticket Jam Occurred in the Lower Level" is described with RDF triples, linked with three kinds of ontologies: device, device in trouble and alarm ontologies. Trouble cases from working reports are mainly described with the property set from device in trouble ontology: working hours, branch office name (= the place of trouble occurrence), cause (for trouble case), manufacturing name (for the device), trouble handling method and similar trouble. So far, the value of trouble handling method and similar trouble, are juts literals but should be structured with a set of other properties for further knowledge retrieval.

4 System Overview

This section describes the preprocessing for trouble handling methods written in work reports and how to specify three types of domain ontologies and build them on a server. Then we develop a web application to enable us to acquire information related to trouble handling methods with RDF triples using domain ontologies by RDF query language endpoint.

4.1 Natural Language Processing

In this section, we explain the details of the natural language preprocessing we performed before generating the domain ontology from work reports save as Excel files.

Dictionary Registration. In the work, there are several technical terms of both the work and devices. The majority of the technical terms are combined words of multiple morphemes, and we registered these technical terms using the dictionary function in a morphological analysis library for Python, without breaking them down into multiple words in the morphological analysis.

Correction for Orthographical Variants. We corrected the small differences in expressions that can occur when different people input data.

Removal of Unnecessary Information. In the comments of work reports, sometimes the date and time of the occurrence of the trouble of devices or an equipment number may be described; however, there are independent items that describe these pieces of information. Hence, those pieces of information are not important in the analysis process and were eliminated them from the analysis.

Extraction of Alarm Information. Information about what type of alarm occurs is sometimes written in the comments, we extracted the alarm information from them.

By considering a series of the aforementioned four processes above as one cycle, we processed the data by performing multiple cycles.

Calculating the Degree of Similarity Between Trouble Handling Method. Based on the assumption that, when referring to the trouble handling method for a certain trouble of devices, we can acquire more information if similar trouble handling method are provided, we regarded each of the trouble handling method as a document and calculated the degree of similarity between the documents. For example, the meaning of two sentences "After confirming A, I perform B" and "After performing X, I confirm A and perform B" are similar but the latter provides more information.

When calculating a document vector, we applied a weighting factor based on term frequency–inverse document frequency (TFIDF). If we calculate a document vector simply based on the frequency of its appearance, the influence of the words such as "confirmation" and "recovery," which not only appear in particular documents but also broadly appear in several general documents, is not excluded. Hence, using TFIDF, we applied weighting factors to the words that characterize the document. We then defined the degree of similarity by the inner product of the two document vectors. This means that if the calculated value is close to unity, the degree of similarity is high.

We then calculated the degree of similarity between all the trouble handling method. We defined the "similar trouble handling method" as the trouble handling method having a degree of similarity of 0.6 or greater and that is within the top five degrees of similarity, even though this is a provisional judgement.

4.2 Constructing Domain Ontology

After completion of natural language processing, we created three types of domain ontology (device ontology, device in trouble ontology, alarm ontology) described in Sect. 3.2 by using Apache Jena [7], which is a Java library for building linked data.

About device in trouble ontology, we defined an object property that represents the similarity of the trouble handling method, and we created an instance–property network based on the degree of similarity in trouble handling method and the definition of similar trouble handling method mentioned in the previous section.

Finally, we defined an object property "abnormality," and using this property, we created an instance–property network between the instances in the three types of ontology.

4.3 Application

We developed an application that acquires information related to trouble of devices from web ontology language files (the domain ontology) placed on a server through a SPARQL endpoint. In this section, we explain three functions of the application.

Displaying Classes. First, the classes of alarms, devices, and trouble of devices are displayed on the left-side of the application display. If we select an upper class, the class that is one level lower will be expanded, and if we select the lowest class, its contents will be reflected as the input value in the search box.

Alarms are categorized by alarm-generating instruments, and we can effectively select the alarm type from the alarm class. Furthermore, if an abnormality occurs even when no alarm sounds, we can select a device that may be related to the trouble of devices (Fig. 2).

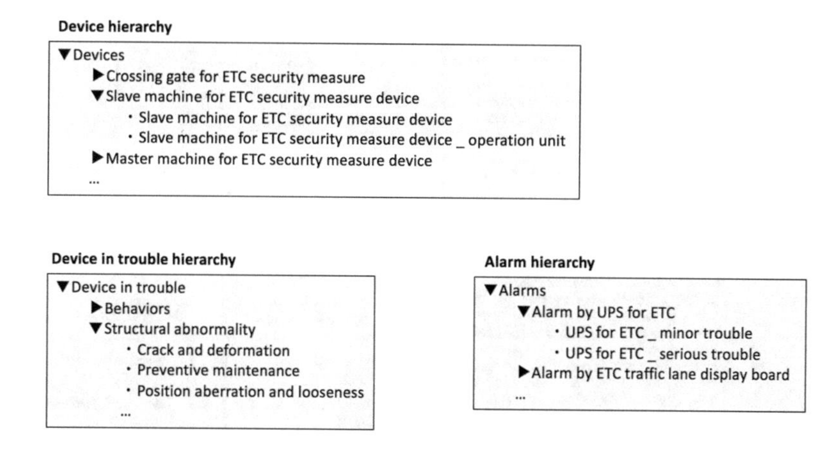

Fig. 2. Displaying classes.

Search. If we select the lowest class, its contents are reflected as the input value for the search box. If we select the search button, a list of related data of devices in trouble is displayed in a table format through the instance–property network of the "abnormality" property.

In addition, if we enter words into the search box, we can acquire a list of data of devices in trouble that contains trouble overviews that include those words (Fig. 3).

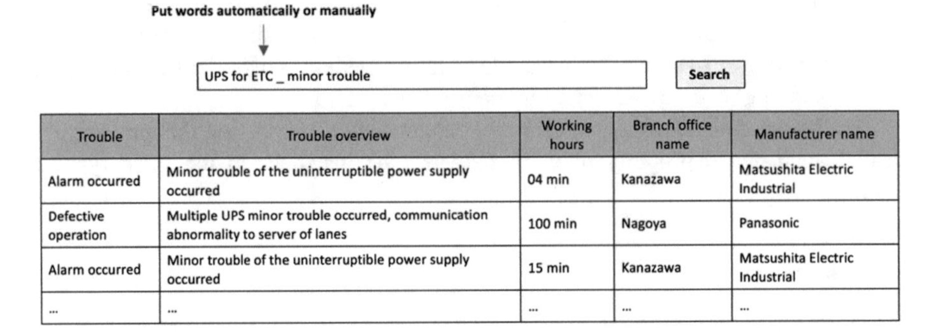

Trouble	Trouble overview	Working hours	Branch office name	Manufacturer name
Alarm occurred	Minor trouble of the uninterruptible power supply occurred	04 min	Kanazawa	Matsushita Electric Industrial
Defective operation	Multiple UPS minor trouble occurred, communication abnormality to server of lanes	100 min	Nagoya	Panasonic
Alarm occurred	Minor trouble of the uninterruptible power supply occurred	15 min	Kanazawa	Matsushita Electric Industrial
...	...	...	...	...

Fig. 3. List of search results.

Acquiring Detailed Information. On selecting a row in the list of the data of devices in trouble, the device name, instrument name, cause, and trouble handling method are displayed. In addition, we can display up to five data of devices in trouble if treatments similar to the trouble handling method were performed. Two display formats are available for the data of devices in trouble wherein similar treatments were performed: a table format, which displays the items quickly, and a graph format, which aids in visualizing the instance–property network (Fig. 4).

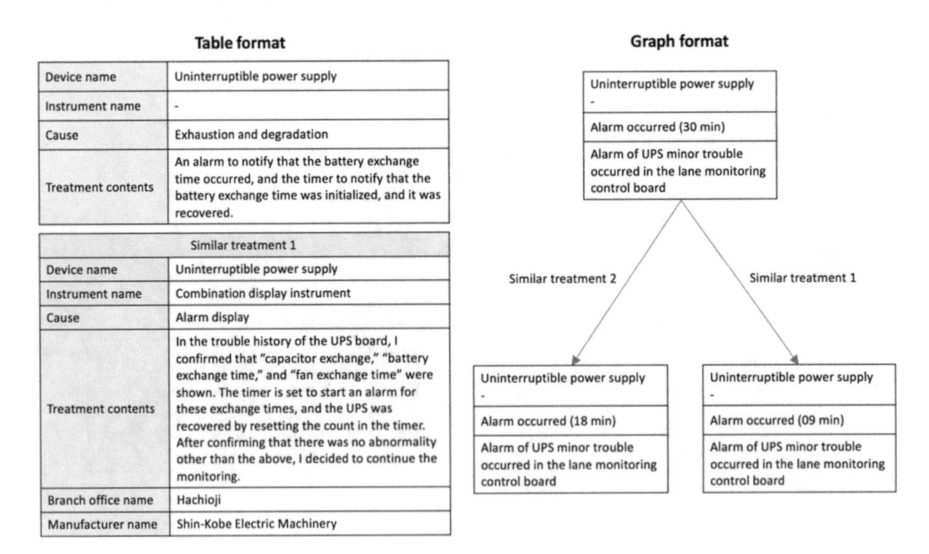

Table format

Device name	Uninterruptible power supply
Instrument name	-
Cause	Exhaustion and degradation
Treatment contents	An alarm to notify that the battery exchange time occurred, and the timer to notify that the battery exchange time was initialized, and it was recovered.

Similar treatment 1	
Device name	Uninterruptible power supply
Instrument name	Combination display instrument
Cause	Alarm display
Treatment contents	In the trouble history of the UPS board, I confirmed that "capacitor exchange," "battery exchange time," and "fan exchange time" were shown. The timer is set to start an alarm for these exchange times, and the UPS was recovered by resetting the count in the timer. After confirming that there was no abnormality other than the above, I decided to continue the monitoring.
Branch office name	Hachioji
Manufacturer name	Shin-Kobe Electric Machinery

Fig. 4. Acquiring detailed information.

5 Evaluation and Discussion

5.1 Case Study

Case Study Overview. We chose the troubleshooting work on the electronic toll collection (ETC) lane of an expressway maintenance management company as the case study. The objective of the work on ETC lane trouble is to realize the quick recovery of the ETC lane using alarms from the ETC devices. As per the work, ETC inspectors are in charge of the investigation and treatments to deal with trouble of devices, but there is no instruction manual for the work, and skilled ETC inspectors are completing this work based on their own experience and intuition. The work knowledge of the experts is tacit knowledge for each inspector and is not summarized in the documents. Only the reports written after the work is completed contained recorded situations and solutions.

It should be noted that, depending on the alarm type, the subsequent response flow varies, and it is important to properly understand it. Therefore, in the work, an ETC inspector determines alarm type by observing what is displayed on the lane monitoring control board, which is a component of the ETC machine.

Target Data. We used Excel files stored in an expressway maintenance management company as the data for creating the domain ontology, which is the knowledge source. The Excel files had 10,000 rows and 130 columns for the five years of the target interval, and they provide much information regarding the work on ETC lane trouble.

The Excel files contain items that have input formats ("branch office name," "weather," "occurrence date and time," etc.) and items without input formats ("trouble overview," "trouble handling method," etc.). Among these items, "trouble handling method" corresponds to work reports, which describe the trouble situations and performed treatments. We can acquire information regarding actual past examples from those reports, and we cannot deny the usefulness of this item from the viewpoint of sharing work knowledge. However, at present, ETC inspectors do not have sufficient opportunities to read though past trouble handling method.

Incidentally, through the cycle of natural language processing mentioned in previous section, we acquired a total of 58 types of alarm information from 1,700 data out of 10,000 data available.

5.2 Experiment Overview

Two employees (who had experience of the target work) from an expressway maintenance management company participated in this study. Using the application of the proposed system, they addressed two quizzes of constructing workflow based on the information provided by the application, and we conducted a questionnaire survey.

5.3 Experimental Results

Opinions of Persons in Charge of the Work. We acquired opinions regarding the usefulness of the proposed system from the viewpoint of the persons in charge of the work.

Appraisable points:

- The alarm classes are useful because we can quickly identify the correct alarm.
- The device classes are also useful because it is sometimes necessary to check a trouble example based on a device.
- It may be possible to use the class structure to estimate which part is out of order.
- It may be possible to acquire useful information from the data of devices in trouble in which similar treatments were performed.
- It is possible to refer to data of devices in trouble acquired from other offices.
- It is useful to have access to similar cases.
- The attempt to reuse work reports is interesting.

Problems:

- Information regarding the detailed procedures and the basis of the trouble handling method are missing.
- There are too few trouble examples available.
- As multiple people have input the original data, there are variations in the level of details of the provided information.
- There are time constraints in the work, and after an alarm occurs, there is insufficient time to carefully inspect the past data of devices in trouble by using the system.

Usefulness as Support for Constructing Workflow. We examined the usefulness of the information provided by the application of the proposed system as support for constructing a workflow for the work.

In the process of constructing a workflow, the first quiz was to create a workflow to handle the alarm “abnormality in the control processing part in traffic lane server.” It was difficult to solve this problem because of the large number of branches, and it was difficult to construct a workflow based only on the information provided by the application. In contrast, the second quiz was to construct a workflow to handle the alarm “abnormality in the storing part in traffic lane server,” and there were a small number of branches. Thus, a simple workflow was constructed in approximately 1 h by extracting a series of flows of work from the information provided by the application.

Hours Required to Develop the System. We recorded the number of hours consumed for construction of the knowledge expression in this study (Table 1).

In this study, approximately 47 h were required for the preprocessing of the data. Only approximately 30 min were required for the construction of the domain ontology after the preprocessing because this process was automated. Furthermore, after carrying out the workflow construction experiments, we found that constructing a workflow for one alarm type required 1 h, which means that we required approximately 58 h to construct workflows for all the 58 types of alarms. Hence, we required approximately 105 h to construct the knowledge expression.

Table 1. Hours required for system development

Hours required to construct knowledge expression	Approximately 105 h
Degree of detail in a workflow	Coarse
Application development period	Approximately 2 months
Corresponding alarm number	58 types

5.4 Discussion

Discussion Regarding the Usefulness of the Application for Business Support. The function by which we can quickly select an element by the element classification was in line with the work-related demand of the work site. In addition, the function was useful in providing data of devices in trouble in which similar treatments were performed. However, there are some problems such as an insufficient number of work report data, missing detailed information, and time constraints at the work site, and we found that it is difficult to use the present system in its current state to support the work. Nevertheless, we received a favorable evaluation for the attempt to reuse the work reports.

Discussion Regarding System Development Hours. The system gives a structure to the information of work reports and the application effectively provides business knowledge, which enabled the system to provide support for the construction of many simple workflows to handle a great number of alarms. Thus, if the system assist constructing simple workflows and then employees refine the workflows based on interviews with experts, we expect that the cost for constructing knowledge expression would be reduced as compared to a case wherein workflows prepared from scratch.

6 Conclusion

In this study, we have developed and evaluated a support system for acquiring knowledge by giving a structure to the information of work reports.

The results confirmed that the system provide support for the construction of simple workflows to handle a great number of alarms. We found out that usefulness of giving a structure to the information of work reports.

Furthermore, according to the opinion from workers, it can be said that the proposed system can more quickly search for data of devices in trouble the user wants to know than traditional work reports because of utilization of domain ontology and instance–property networks.

However, we found that there are some problems such as the insufficient number of work report data, missing detailed information, and insufficient time for using the system when dealing with the work site, and the usefulness of the proposed system for business support was inhibited by these issues. Therefore, we consider applying the proposed system to other businesses in which there are fewer constraints with respect to the quality and quantity of available data and the situations at the work sites in order to demonstrate the availability to work sites.

References

1. Takeuchi, N.: Knowledge management and text mining. In: Encyclopedia of Artificial Intelligence, pp. 1347–1348. Kyoritsu Shuppan Co., Ltd., Japan (2017)
2. Razmerita, L., Angehrn, A., Maedche, A.: Ontology-based user modeling for knowledge management systems. In: The 9th International Conference, UM 2003, Johnstown, PA, USA, pp. 213–217 (2003)
3. Ido, T., Kitamura, Y.: Constructing knowledge models and an ontology for adjusting specifications of industrial machinery. In: The 32nd Annual Conference of the Japanese Society for Artificial Intelligence, Japan (2018)
4. Athira, M., Sreeja, M., Reghuraj, P.: Architecture of an ontology-based domain-specific natural language question answering system. Int. J. Web Semant. Technol. **4**, 31–39 (2013)
5. Kitamura, T., Mizoguchi, R.: Ontology. In: Encyclopedia of Artificial Intelligence, pp. 1277–1280. Kyoritsu Shuppan Co., Ltd., Japan (2017)
6. Kosaka, K.: Domain ontology developments and their utilization. J. Jpn. Soc. Inf. Knowl. **19**(4), 296–305 (2009)
7. Apache Jena. https://jena.apache.org/

Support Tool for Refining Conceptual Model in Collaborative Learning

Misaki Maruyama, Shinpei Ogata(✉), Kozo Okano, and Mizue Kayama

Shinshu University, Nagano-shi, Nagano 380-8553, Japan
{ogata,okano,kayama}@cs.shinshu-u.ac.jp

Abstract. Conceptual modeling using the UML class diagram notation is one of the essential tasks in object-oriented analysis. In collaborative learning, it is difficult for learners to properly share the task of conceptual modeling in their group because a few members who are more skilled out of their group are apt to refine and maintain their single conceptual model. Consequently, as a result of iterative refinement, the others easily become unable to recognize links between their single conceptual model and other documents such as use case descriptions. To improve this problem, this paper proposes a support tool for refining a conceptual model in collaborative learning. This tool enables to automatically merge multiple partial conceptual models, to semi-automatically refine elements of the merged model, and to automatically reflect the refined elements in each partial conceptual model. The effectiveness of this tool was evaluated by applying it to a course of object-oriented development.

Keywords: Class diagram · Collaborative learning
Conceptual modeling · Object-oriented analysis

1 Introduction

To properly analyze and design object structure (e.g. object's classes, attributes, associations and operations) is one of the essential tasks in object-oriented development. A UML class diagram emphasizing data structure (e.g. object's classes, attributes and associations) of a target domain is often called a conceptual model [1,2] (or a domain model [3,4]). Many methods to obtain a conceptual model by deriving it from the corresponding use case model have been proposed [3–5]. Especially, Boronat et al. presented an approach for obtaining a conceptual model by merging partial conceptual models which are derived from each use case in order to keep their traceability [5].

In collaborative learning, a few members who are more skilled out of their group members are apt to refine and maintain their single conceptual model. Consequently, as a result of iterative refinement, the others easily become unable to recognize links between their single conceptual model and other documents such as use case descriptions. In such situation, *"free-rider effect where one team member just leaves it to the others to complete the task"* [10] is easy to happen.

M. Virvou et al. (Eds.): JCKBSE 2018, SIST 108, pp. 147–157, 2019.
https://doi.org/10.1007/978-3-319-97679-2_15

Hence, where each member was assigned a task to analyze one or more use cases, Boronat's approach is expected to be reasonable for preventing the *free-rider effect*. However, this approach also has two disadvantages by compared with handling a single conceptual model. One hand is that it is difficult for all members to grasp their entire structure since a conceptual model is distributed as partial models. The other hand is that there are laboring tasks such as to merge partial conceptual models for the review of the entire structure, to reflect the reviewed model in each partial model. Although a method proposed by Boronat et al. [5] enables to merge multiple models, it seems not to directly reflect a refined model in the corresponding partial models.

Some methods to facilitate collaborative design learning [6–8] and distributed collaborative design [8,9] have been proposed. However, there is not an enough tool satisfying the following requirements in order to solve the problems mentioned above.

1. Each member can keep to maintain a partial conceptual model for each use case.
2. A tool can support automatic merging and reflection of conceptual models so that learners can continually maintain partial conceptual models and iteratively review and refine the merged conceptual model.
3. Instructors and learners who may be novices at using a tool can easily introduce and use the tool.

To establish a tool satisfying requirements mentioned above, this paper proposes a support tool for refining a conceptual model in collaborative learning. This tool enables to automatically merge multiple partial conceptual models, to semi-automatically refine elements of the merged model, and to automatically reflect the refined elements in each partial conceptual model.

The effectiveness of this tool was evaluated by applying its prototype tool to a course of object-oriented development and by comparing the result with the result of applying a more manual way. As a result, the proposed tool compared with a more manual way enhanced the consistencies among partial conceptual models and the merged conceptual model, and the efficiency of maintaining such conceptual models.

2 Support Tool for Refining Conceptual Model

2.1 Overview

In this section, we describe the overview of the proposed tool for refining conceptual model in collaborative learning. Figure 1 shows the overview of steps to use our proposed tool. The proposed approach using the tool consists of four steps: (1) preparing use case documents and the corresponding partial conceptual models, (2) merging partial conceptual models int a conceptual model, (3) refining the conceptual model, and (4) reflecting the.refinements in each partial conceptual model. Steps 2 and 4 are fully automated and step 3 is partly automated.

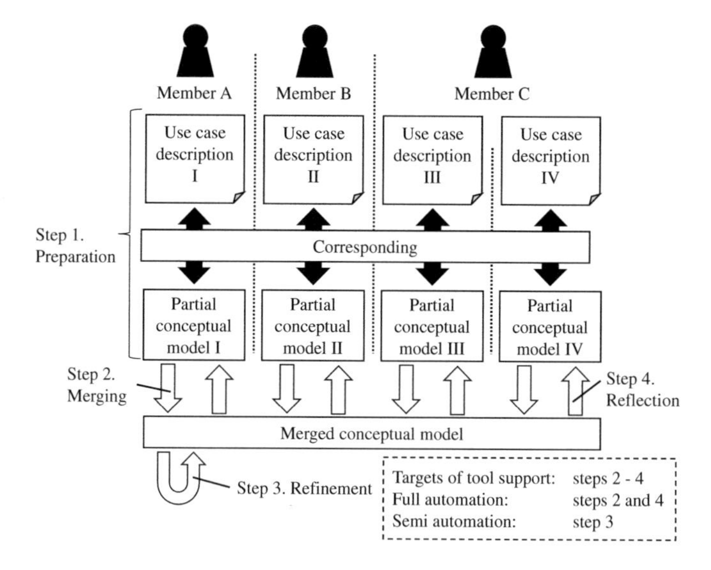

Fig. 1. Overview of steps to use proposed tool

At step 1, members define use cases, and then divide the tasks of writing use case descriptions among them. As an option for reducing the variety of naming an element, they create terminology used in target domains before writing the descriptions. Then, each member writes his/her use case descriptions and creates a partial conceptual model for each of those descriptions.

At step 2, the proposed tool receives two or more partial conceptual models as the inputs, and generates a single merged conceptual model automatically. The detail of this step is described in Sect. 2.2.

At step 3, the members review the merged conceptual model, and refine it by using an editor, which is a part of the proposed tool. This review and refinement are performed in non-distributed and synchronous learning environment so that learners can concentrate their effort on solving problems on modeling. The reason why the learners need such environment is because they do not usually have modeling skills and domain knowledge so much, and need serious discussions about target domains and their models. To support distributed or asynchronous learning environment is more practical, but such environment is easy to make a lot of problems on their communication rather than on modeling. The detail of how to refine a conceptual model in this step is described in Sect. 2.3.

At step 4, the refined conceptual model at step 3 are reflected in each original partial conceptual model inputted at step 2. This means that the proposed tool outputs multiple partial conceptual models by receiving one refined conceptual model as the input. The detail of this step is described in Sect. 2.4.

2.2 Merging Partial Conceptual Models

When partial conceptual models are merged, equality between their elements has to be calculated to determine the operation (i.e., merging or combining).

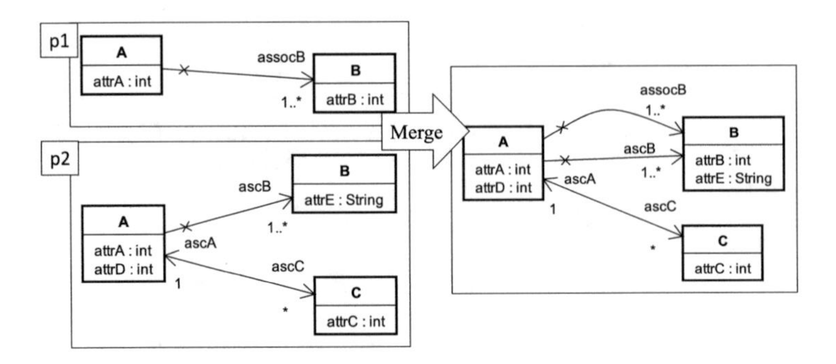

Fig. 2. Example of merging partial conceptual models

For such calculation, we determined an identifier of each element type (i.e. a class, an attribute and an association) as shown in Table 1. It notes that the type "Association-End" is dealt with as a part of the type "Association". In addition, an association is captured every direction (i.e. A to B and B to A) in this research. Thus, the identifiers of associations are different between the directions even if two associations are represented as the same line segment. This means that an association is captured every association-end.

The merging algorithm we adopted is simple so that learners can easily trace their model elements between partial conceptual models and merged one. In this algorithm, all partial conceptual models are integrated into one conceptual model. If the identifier of an elements is completely same to the identifier of another element, then those elements are merged to one element. If not then both those elements present in the merged model. Regarding the order of merging, attribute and association elements are merged after class elements were merged.

Such the simple algorithm may be unsuitable for large systems providing with many use cases because it may not reduce model elements enough to keep the readability. Conversely, we believe that such the algorithm is proper in the context of education in which learners are novices and the size of a model will be not large.

Figure 2 shows an example of merging partial conceptual models. The right model is the result of merging the left two models, i.e. p1 and p2. According to Table 1, both the identifiers of the attributes "attrA" included in both the left models are equivalent to "A_attrA_int". Thus, they were merged to one element as the attribute "attrA" in the right model. Meanwhile, regarding only the name that is a part of the identifier of "Association", the association-end "assocB" is different from "ascB". Thus, both these elements appears in the right model.

When model elements are merged, their source models and identifiers are recorded onto the resulting merged model. This means that each model element owns the source models and identifier as properties. Source models mean the names of the partial conceptual models respectively owning the element named by the same identifier.

Table 1. Identifier of each element type

Element type	Identifier
Class	*clsName*
Attribute	*clsName_attrName_attrType*
Association	*srcAssocEnd_assocName_tgtAssocEnd*
Association-End	*srcName_tgtName_endName_multiplicity_nav_aggrType*

The "*clsName*" means a class name. The "*attrName*" means an attribute name. The "*attrType*" means an attribute type. The "*srcAssocEnd*" and "*tgtAssocEnd*" are replaced to the identifier of "Association-End" respectively. The "*assocName*" means an association name. The "*srcName*" means the source class name (i.e. owner name) of an association-end and the "*tgtName*" means the corresponding target class name (i.e. type name) of the association-end. The "*endName*" means an association-end name. The "*multiplicity*" means the multiplicity (e.g. `0..1`, `1`, `*` or `1..*`) of an association-end. The "*nav*" means the navigability (i.e. `non_navigable`, `navigable` or `unspecified`) of an association-end. The "*aggrType*" means the aggregation type (i.e. `none`, `aggregation` or `composition`) of an association-end.

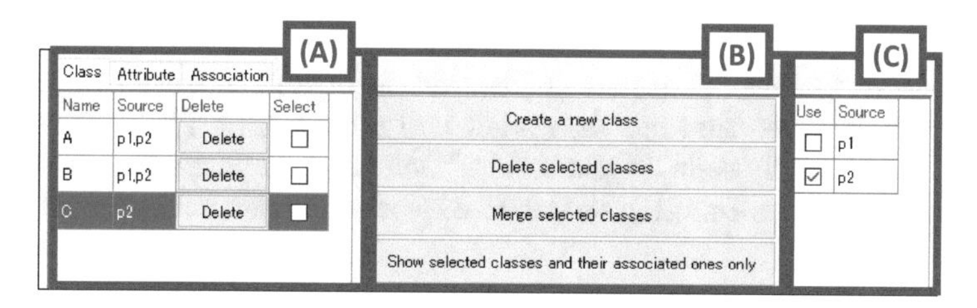

Fig. 3. Class editor view provided by proposed tool

2.3 Refining Merged Conceptual Model

A conceptual model simply merged needs many operation to refine it since it may contain many unnecessary elements and/or multiple sets of the elements that have different name but the same meaning. The creations or deletions of elements, or the modification of a few properties may be easy operations on a graphical editor such as astah [11] for some learners. However, the manual merging of similar elements is often laboring and untraceable work since multiple operations of creations, modification and/or deletion will be combined. Therefore, the proposed tool supports actions of merging, creation, modification, and deletion of elements to enhance these actions easier and more traceable. In addition, the proposed tool also supports an action of controlling the view of a merged conceptual model containing many classes so that learners can easily focus on elements concerned with. All learners of a group have to participate in the review and refinements of their merged conceptual model through a discussion. In parallel with the review and refinements, one of the learners of the group uses the proposed tool to refine the model on the basis of the discussion.

Figure 3 shows the class editor view provided by the proposed tool, which deals with the right model of Fig. 2. Part (A) shows a table-base property editor in which learners can edit any properties relating the identifier of the corresponding element type. Part (B) shows control actions regarding classes. Especially, the actions "Merge ..." and "Show ..." are important. The action "Merge ..." is executed when the elements selected in the table-based editor are merged. In the process of merging, learners input and determine a value of each property such as class name because the multiple different candidates of the value are extracted among the selected elements and it is normally difficult to automatically determine the most proper value from them. For attributes and associations as well, learners determine the properties composing their respective identifier. The action "Show ..." is executed when learners want to confirm only the elements concerned with. The editor shows the elements selected in the table-base editor and their associations and opposite classes only.

Part (C) shows a list of source models. When a row in the table-based editor is focused, the cell of "Source" in the row is reflected in the corresponding check boxes in the list. When a cell of "Use" in part (C) was checked, the corresponding source model name is automatically added to the corresponding cell in part (A) with the delimiter ",". This property is used to determine which partial conceptual models must have which model elements, at step 4.

The action "Create ..." in part (B) is used, for instance, when overlooked elements were discovered through the review and refinements. The action "Delete" in part (A) is basically same to the action "Delete ..." in part (B), but the action in part (B) provides a function to delete one or more elements at once.

2.4 Reflecting the Refinements in Each Partial Model

After such refinements using the editor, the refined conceptual model is reflected in the partial conceptual models inputted at step 2. The identifiers recorded via the merging at step 2 are never changed at step 3. Thus, the operations (i.e. `create`, `update` and `delete`) for reflection can be determined by comparing the current identifiers of refined model elements with the recorded identifiers.

2.5 Implementation

We implemented the above features as a plugin for astah [11], which is a UML modeling tool having plugin development kit for Java. Hence, we have prototyped a tool consisting of two applications: a GUI application to support steps 2 and 4 by using the java APIs; an astah plugin application to support step 3.

3 Case Study

3.1 Overview

We applied the prototype tool (hereafter, called the tool) into a course of object-oriented development. 24 students registered this course and formed 6 groups of

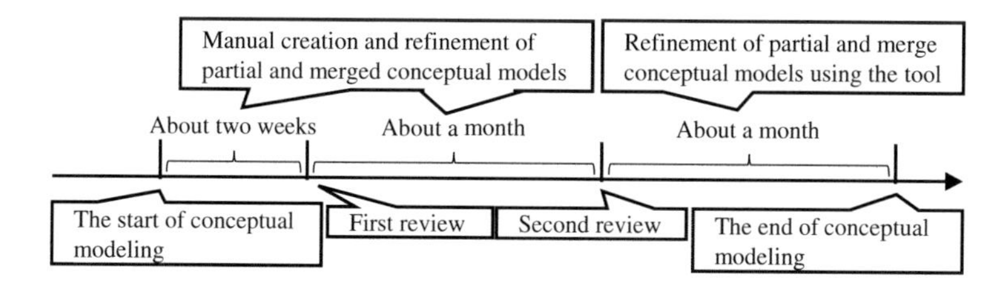

Fig. 4. Schedule of conceptual modeling

4 members. They had fundamental knowledge of Java but little of UML before starting this course. Each group determined a theme of an Android application and then developed it through PBL (Project-Based Learning).

Figure 4 shows the schedule of conceptual modeling. The instructor instructed every group to create a partial conceptual model for each use case and then to merge all those partial conceptual models. Each group manually created and refined conceptual models for a period of about 1.5 months. Then, each group refined the conceptual models using the tool for a period of 1 month.

For evaluating the tool, we investigated the resulting models and students' opinions. In the investigation of the models, the models at the Second Review (SR) and at the End of conceptual Modeling (EM) were used. The following steps are a summary of how to measure inconsistency among partial conceptual models for each phase (i.e. SR and EM).

1. We merged partial conceptual models by using the tool.
2. We identified elements which have the same meaning in the automatic merged conceptual model by referring to the manual one. When multiple elements which have the same meaning but not the same name to each other were found, we counted them. This means that we counted the number of the sets containing two or more such elements.
3. We summed up the numbers counted for every group.

Meanwhile, the following steps are a summary of how to measure inconsistency between partial conceptual models and the corresponding merged conceptual model for each phase (i.e. SR and EM).

1. We merged partial conceptual models by using the tool.
2. We compared the automatic merged conceptual model with the manual one. When elements which appeared only in either of those models were found or when elements which have the same meaning but not the same name to each other between those models were found, we counted them. This means that we counted the number of the sets containing one or more such elements.
3. We summed up the numbers counted for every group.

In these steps, we dealt with all classes and their attributes and associations needed in each target domain, but not the classes that are not represent domain-specific conceptions (e.g. `String` and `List`).

Table 2. Result of measuring conceptual models

Element type	Ics. qty (SR)	Total qty (SR)	Ics. qty (EM)	Total qty (EM)
Set of classes (P-P)	8	60	0	59
Set of attributes (P-P)	2	82	4	91
Set of associations (P-P)	6	88	0	85
Set of classes (P-M)	13	122	1	118
Set of attributes (P-M)	10	166	6	182
Set of associations (P-M)	61	181	5	165

The "Ics. qty" means the qty of the sets determined as inconsistent. The "P-P" means "among partial conceptual models". The "P-M" means "between partial conceptual models and the corresponding merged conceptual model".

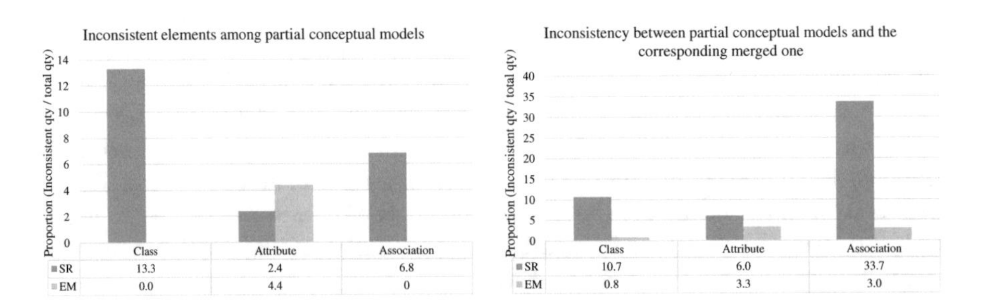

Fig. 5. Ratio of total qty to inconsistent qty

Meanwhile, in the investigation of the students' opinions, we created and used a questionnaire regarding the tool. A summary of those questions are as follows: did each function (i.e. merging, refinement, and reflection) enhance its efficiency?; were all the functions comprehensively useful?; why did you think so?; which of these functions was the best comparatively?

3.2 Result

Table 2 shows the result of measuring conceptual models and Fig. 5 shows each ratio of total quantity (qty) to the corresponding inconsistent qty. Overall, the tool reduced inconsistent element sets of the models manually handled.

Figure 6 shows the result of a part of the questionnaire. According to the result, the answers accounting for over 60% to each function indicate that the function enhanced the efficiency of the corresponding step. In addition, the answers accounting for over 70% indicate that the tool was comprehensively useful.

A summary of positive opinions was as follows: models were easily merged and reflected by a few operations; the operations to merging, refinement and reflection were made more efficient than manual ones; the inconsistency between

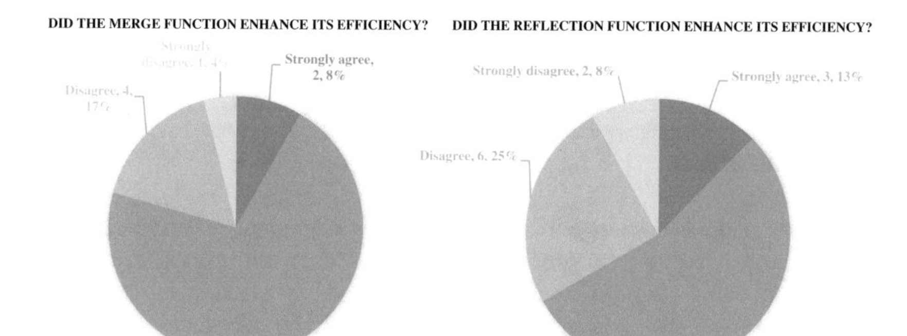

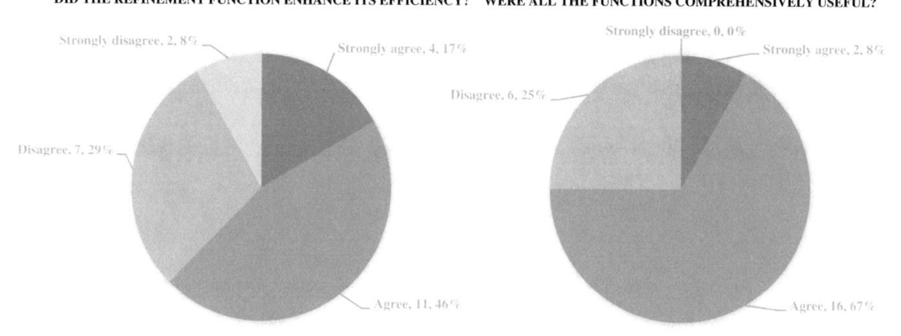

Fig. 6. Result of a part of questionnaire

models became easy to discover; the function to control the view of class diagrams was useful. Conversely, a summary of negative opinions was as follows: difficult to use or boot; laboring work to adequately arrange model elements on a diagram; there were cases that functions behaved unexpectedly; the tool effectiveness was low when classes was not many.

3.3 Consideration

Overall, the inconsistency of the conceptual models at the period of refining them using the prototype tool was improved by comparing with the models manually refined. Therefore, the proposed tool was effective to improve such the inconsistency. However, some inconsistencies that the prototype tool could not improved were caused by tool's bugs or misuses. Hence, the effectiveness of the proposed tool is expected to be enhanced by improving the prototype tool. Meanwhile, the determination of the inconsistency was performed by the experimenter. Hence, the traces between conceptual models should be precisely recorded in future evaluation.

Many students evaluated the effectiveness of the prototype tool positive from the aspect of the efficiency. Hence, the proposed tool was effective to enhance such the efficiency. However, especially arranging model elements on a diagram is needed whenever a merged conceptual model was opened for the first time. We plan to improve this problem according to existing point of view [12].

4 Conclusion

This paper proposed a support tool for refining a conceptual model in collaborative learning. This tool supports automatic merging and reflection of conceptual models, and semi-automatic refinement. Through the evaluation, we obtained an expectation that the proposed tool can enhance the efficiency of collaborative learning of conceptual modeling.

As future work, we will implement a function to get changelogs of model elements in order to grasp the model changes and analyze the learners' tendencies to make mistakes in the refinement step. In addition, we improve the layout of elements on a class diagram by using existing layout algorithms.

Acknowledgment. This work was supported by JSPS KAKENHI Grant Numbers JP16H03074.

References

1. Trujillo, J., Palomar, M., Gomez, J., Song, I.-Y.: Designing data warehouses with OO conceptual models. Computer **34**(12), 66–75 (2001)
2. Koch, N., Kraus, A.: The expressive power of UML-based web engineering. In: Proceedings of the 2nd IWWOST 2002, pp. 105–119 (2002)
3. Thakur, J.S., Gupta, A.: AnModeler: a tool for generating domain models from textual specifications. In: Proceedings of the 31st IEEE/ACM International Conference on Automated Software Engineering (ASE 2016), pp. 828–833 (2016)
4. Fortuna, M.H., Werner, C.M.L., Borges, M.R.S.: Info cases: integrating use cases and domain models. In: Proceedings of the 16th IEEE International Requirements Engineering Conference, pp. 81–84 (2008)
5. Boronat, A., et al.: Formal model merging applied to class diagram integration. Electron. Notes Theor. Comput. Sci. **166**, 5–26 (2007)
6. Hansen, K.M., Ratzer, A.V.: Tool support for collaborative teaching and learning of object-oriented modeling. In: Proceedings of the 7th Annual Conference on Innovation and Technology in Computer Science Education (ITiCSE 2002), pp. 146–150 (2002)
7. Asensio, J.I., et al.: Collaborative learning patterns: assisting the development of component-based CSCL applications. In: Proceedings of the 12th Euromicro Conference on Parallel, Distributed and Network-Based Processing, pp. 218–224 (2004)
8. Chen, W., et al.: CoLeMo: a collaborative learning environment for UML modelling. Interact. Learn. Environ. **14**(3), 233–249 (2006)
9. Mehra, A., et al.: A generic approach to supporting diagram differencing and merging for collaborative design. In: Proceedings of the 20th IEEE/ACM International Conference on Automated Software Engineering (ASE 2005), pp. 204–213 (2005)
10. Kerr, N.L., Bruun, S.E.: Dispensability of member effort and group motivation losses: free-rider effects. J. Pers. Soc. Psychol. **44**(1), 78–94 (1983)

11. Change Vision, Inc.: Astah professional. http://astah.net/. Accessed 21 Mar 2018
12. Nikiforova, O., et al.: Several issues on the layout of the UML sequence and class diagram. In: Proceedings of the 9th International Conference on Software Engineering Advances (ICSEA 2014), pp. 12–16 (2014)

Tool to Automatically Generate a Screen Transition Model Based on a Conceptual Model

Yukiya Yazawa[1(✉)], Shinpei Ogata[1(✉)], Kozo Okano[1], Haruhiko Kaiya[2], and Hironori Washizaki[3]

[1] Shinshu University, 4-17-1 Wakasato, Nagano-shi, Nagano 380-0928, Japan
17w2094c@shinshu-u.ac.jp, {ogata,okano}@cs.shinshu-u.ac.jp
[2] Kanagawa University, 2946 Tsuchiya, Hiratsuka-shi, Kanagawa 259-1293, Japan
kaiya@kanagawa-u.ac.jp
[3] Waseda University, 3-4-1 Okubo, Shinjuku-ku, Tokyo 169-8555, Japan
washizaki@waseda.jp

Abstract. A screen transition model (STM) is one of the effective design models to specify user interface structure and behavior. Moreover, a lot of methods to generate a prototype system from a STM have been proposed. However, developers are still required to create complex STMs manually and ensure that the STMs are consistent with other models. Therefore, we propose a tool to automatically generate a STM from Object CRUD (OCRUD) diagrams. OCRUD diagrams can be obtained by concreting a conceptual model which represents the data structure of domains. Evaluation results show that the proposed tool was useful for STM modeling and discovering mistakes such as omissions of elements to model.

Keywords: Modeling support · Screen transition model
Object CRUD diagram · Model driven development

1 Introduction

Model driven development (MDD) [1] is one of the efficient development methods of software. In MDD, use case models [2], conceptual models [1], activity diagrams [3], etc. are created and source codes are automatically generated based on those models. Generally, there is an advantage that it is possible to automatically/semi-automatically generate source codes from models which are easier to understand than source codes. But there is a disadvantage that if there were omissions of description which are missed by developers in those models, the omissions will be reflected in the source code. In order to prevent this disadvantage, it is important for developers to accurately write models without omissions.

M. Virvou et al. (Eds.): JCKBSE 2018, SIST 108, pp. 158–167, 2019.
https://doi.org/10.1007/978-3-319-97679-2_16

1.1 Problem

A STM is a part of the design models in software development and expresses input or output items and the screen transitions between screens. A STM has superior effectiveness in MDD because a prototype system can be generated automatically by a tool [4] which is to generate from a STM. However, developers should write a STM manually and consistency between other models. It is difficult for developers that they write a STM because the complexity of models increases by expanding the scale of software. To the best of our knowledge, there is not any support for creating a STM. Therefore, we think that it is necessary to support for modeling focused on a STM.

1.2 Purpose

We propose a tool to automatically generate a STM from object CRUD diagrams [5] in order to support modeling focused on STMs for beginners of software development. A STM generated by this tool includes information of a conceptual model for achieving use cases of software. Hence, it is considered that the developers can decrease omissions and maintain consistency between a STM and a conceptual model. Also, it can be considered to be convertible to webML [6] or DSML [7] by setting certain conversion rules, so it can be applied to the automatically generation of a prototype system. And it can be expected that traceability between models and source codes should be ensured [5].

In this paper, we evaluated how much the tool can support a generation of a STM for developers. As a result, it was shown that our tool was useful for STM modeling and discovers mistakes such as omissions of modeling.

1.3 Composition of Paper

In the following, we describe models which are used by the tool in Sect. 2, and the proposed tool will be explained in Sect. 3. Section 4 is described an evaluation of the tool and its results, and Sect. 5 is described some related researches. The final section provides a summary of this paper.

2 Description of Models

In this section, models which are used in the proposed tool are described. Figures in this paper are listed as an example is based on an interview reservation system between teachers and students.

2.1 Object CRUD (OCRUD) Diagram

OCRUD diagrams [5] show CRUD (Create, Read, Update, Delete) of objects based on a conceptual model for a use case and actor, and its notation is based on the UML object diagram. In an object diagram, the names of a classifier, slots or links of an instance specification are named after the corresponding conceptual model elements. Meanwhile, the types of C, R, U and/or D are given to instance specification names, slot values, or link names. Figure 1 shows an example of OCRUD diagrams. Figure 1(a) is a conceptual model, and the classes enclosed in the square are required in a use case "cancel a reservation". Figure 1(b) is OCRUD diagrams. Objects described in OCRUD diagrams correspond to the classes enclosed in the square in (a).

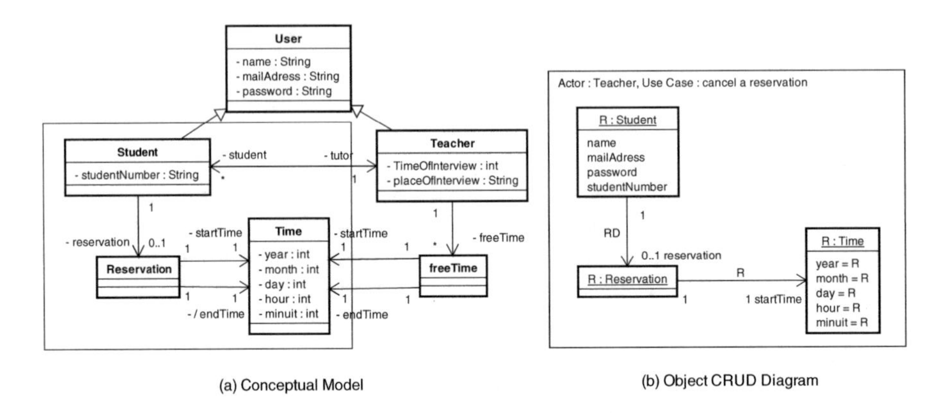

Fig. 1. A conceptual model and an object CURD diagram.

2.2 Screen Transition Model (STM)

A STM [4] is a model representing screen input/output items and transitions between screens, and it includes a screen structure definition and a screen transition definition. In this paper, we refer to them as a screen structure diagram and a screen transition diagram.

Screen Structure Diagram. A screen structure diagram defines screen structures such as input/output items and transitions between screens in a class diagram. An example of a screen structure diagram is shown in Fig. 2. The stereotype of classes in a class diagram is described "boundary" indicating the boundary between users and software. In addition, either "screen" or "component" is inserted in the stereotype. Screen classes represent screens, and component classes represent a part of screens. A type of attributes in the class is specified one of Input, Output, and Link. The Input represents an input item not to cause a screen transition. The Output is an output item such as a character string. The Link represents an input item causing a screen transition. And, it indicates that the corresponding type of components is Input or Output in object units by designating Input or Output at Type in the composition of the component.

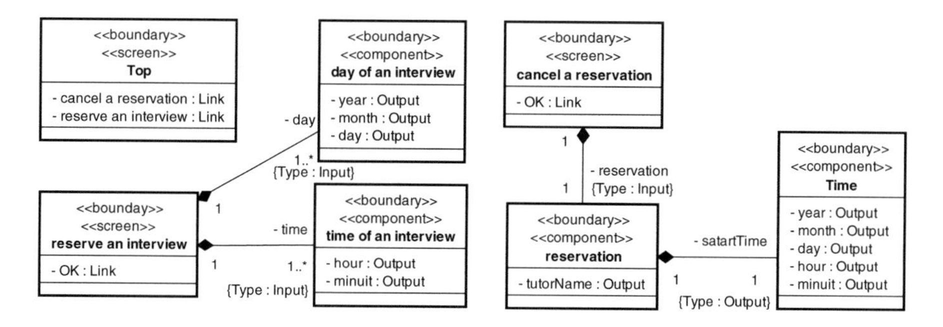

Fig. 2. A screen structure diagram.

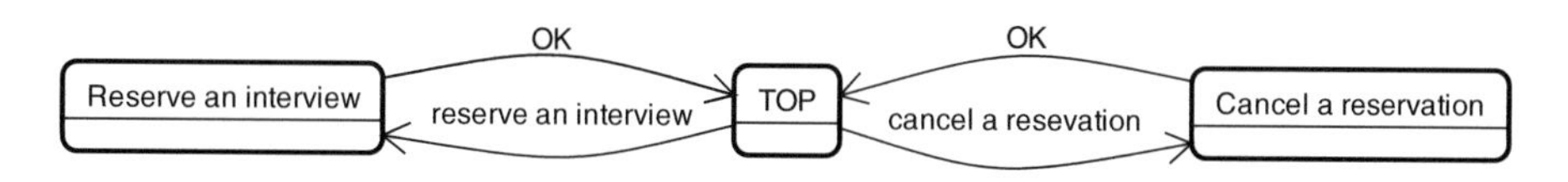

Fig. 3. A screen transition diagram.

Screen Transition Diagram. A screen transition diagram defines transitions between screens by considering states as a screen in a state machine diagram. An example of a screen transition diagram is shown in Fig. 3. The state corresponds to a screen one-on-one, and the transition between states is expressed as a transition between screens. Also, the trigger name is linked to an attribute name of Link which was defined in a screen structure diagram. The direction of the arrow of a transition indicates a destination of a transition of a screen.

3 Proposed Tool

An overview of the proposed tool is shown in Fig. 4. In this tool, OCRUD diagrams is inputted to the model conversion tool, and we can obtain a STM reflecting information of a conceptual model as output. Processes of this tool are as follows:

1. Convert model elements of OCRUD diagrams into model elements of a screen structure diagram in a STM.
2. Link a component class with a screen class in the screen structure diagram.
3. Insert model elements which are generated without the origin in the screen structure diagram.
4. Generate a screen transition diagram in a STM based on the screen structure diagram.

The input to this tool gives an astah [8] file, and the result is output in the form of writing in addition to the same file. The models which are used in this tool are described based on UML [9].

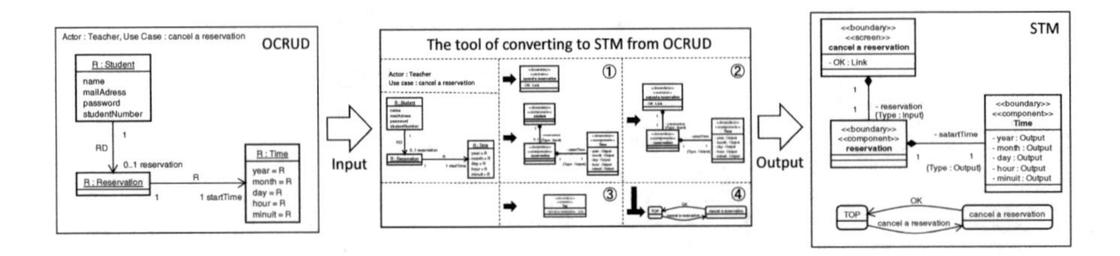

Fig. 4. An overview of the proposed tool.

3.1 Precondition

It is assumed that a conceptual model and OCRUD diagrams have been written as a precondition for automatically generating a STM. OCRUD diagrams are an extension of the existing CURD analysis [10], and there is no difference from a conventional way of thinking.

3.2 Conversion Rule

Table 1 shows the rules for converting OCRUD diagrams to a STM. After OCRUD diagrams is converted to a screen structure diagram, a screen transition diagram is generated based on the screen structure diagram. Table 2 shows model elements that are generated in the conversion process.

Table 1. Correspondence between OCRUD diagrams and screen structure diagram.

Object CRUD diagram	Screen structure diagram
Use case	Screen
Use case name	Class name of screen
Instance specification	Component
Link attribute	Composition
Link name = C, U, D	Type = Input
Link name = R	Type = Output
Link end name	Class name of component
	Associated end name of composition
Multiplicity of Link	Multiplicity of composition
Slot name	Attribute name
Slot value = C, U, D	Type of attribute = Input
Slot value = R	Type of attribute = Output

Table 2. Model elements that are routinely generated in screen structure diagram.

Name of a STM element	Description
Top screen class	Generate screen class whose name is Top in the screen structure diagram
Link to each use case screen from Top screen class	Generate a Link attribute representing the transition to each screen from the Top screen. Name of the attribute is the class name of each screen
Link to Top screen class from each use case screen	Add the Link attribute representing transitions to return to the Top screen when the use case is achieved to each screen excluding the Top screen. Attribute be added is "OK: Link"

Generate Screen Structure Diagram. A screen structure diagram is generated based on Tables 1 and 2. Table 1 shows the correspondence between the elements of OCRUD diagrams and a STM, and a STM is converted from OCRUD diagrams according to Table 1. Firstly, each object in OCRUD diagrams is converted to the component class. In this case, only the associated end name of compositions to a screen class from a component class are named after the classifier name of an instance specification. Next, compositions between a screen class and a component class or between component classes are written based on the link between objects in OCRUD diagrams, and an Associated end name, Multiplicity, and Type are inserted in compositions. Then, if any of C, R, U, D is described the slot in an OCRUD diagrams, it is replaced by the attribute of the screen structure diagram. The name of each attribute is the same name as slot in OCRUD diagrams, and the type of attributes and component classes is converted to Input if it is C, U, D, and Output if it is R. However, it is converted to Input when C/U/D and R are written together. Table 2 is model elements generated routinely without relation to OCRUD diagrams after converting elements in OCRUD diagrams into a screen structure diagram. Finally, if a component class which is linked from a screen class doesn't have attribute, it is erased and its composition is directly linked to the screen class.

Generate Screen Transition Diagram. A screen transition diagram is generated based on the screen structure diagram. At the first, a name of state for each screen in the screen structure diagram is described, and transitions are written to states of each use case screen from a state of a Top screen so as to correspond one-to-one with each screen state excluding a Top screen in the state machine diagram. A name of a Link in a screen class is described in a trigger. Conversely, transitions are written to the state of a Top screen from states of each use case screen and its trigger is described "OK". The screen transition diagram is generated by the above operation.

3.3 Limit of this Tool

In the proposed tool, a STM that conforms to a conceptual model is generated but the layout of a screen in software are not adjusted because the STM is only a model handles input and/or output items and transitions between screens. However, a layout model is possible to be generated from a STM in the existing research [11]. We can designate the layout and the design of the input and/or output items by utilizing it.

4 Evaluation

We applied the proposed tool to 8 system models to evaluate whether we can support or can't support generation of a STM for developers. This evaluation was conducted by comparing a STM which was generated without using the tool (manual STM) and a STM which was generated by the tool (automatic STM). Thereby, it is determined whether the manual STM elements corresponding to a conceptual model were contained in the automatic STM by the proposed tool or not. Then, it is evaluated whether a STM can be generated as the base of the manual STM by the tool.

The target system models in this evaluation are (a) a ATM simulation system, (b) a bulletin board system, (c) an IT words quiz, (d) a payroll system, (e) a reversi and (f) a spot search system. Each of (a)–(f) is a model which is written by a group of three students equivalent to 2nd year of information technology engineering college. They received a review from software engineering experts in generating a STM. An experimenter interpreted CRUD from use case descriptions and a conceptual model, and generated object CRUD diagrams because every model of (a)–(f) was not analyzed from the viewpoint of CRUD.

4.1 Result

The results of the evaluation is shown in Fig. 5. The vertical axis is the number of model elements of a STM (the total number of attributes and components which are related to a conceptual model and type is Input in). Model names of (a)–(f) are described on the horizontal axis. The gray graph represents the total number of model elements of the automatic STM and we regard it as a correct answer number. The black graph represents the number of model elements that were differences between the manual STM and the automatic STM.

4.2 Consideration

From Fig. 5, it is seems that there are no differences between the automatic STM and the manual STM in the three system models. The differences are model elements of the manual STM which were not included some model elements in the automatic STM. Therefore, it can be said that the elements of the automatic STM match model elements which are related to a conceptual model in the

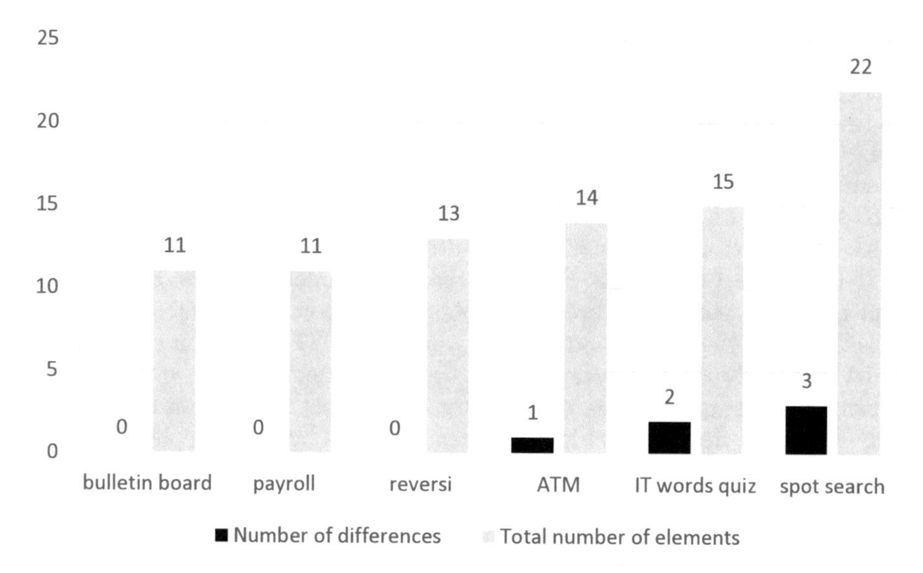

Fig. 5. The result of the evaluation.

manual STM. From this, it was suggested that a STM which is generated by this tool includes information at the base of a STM.

On the other hand, in the other three system models, it can be seen that the automatic STM has model elements which are not included in the manual STM. We considered that those model elements are omissions which were not described when developers wrote a STM because use cases can't be achieved if those model elements are not included in a STM. From this, it is expected that not only the base of the STM can be generated by the proposed tool but also some omissions can be prevented. However, one of the differences in Fig. 5 is not a description omission. The difference is a model element which was aggregated some model elements in the automatic STM into one item. This is regarded as the model element which was aggregated by the idea that it is useful to use the system if some model elements are combined into one. The automatic STM is only the basis of a STM, and it is not considered usability such as the elements combed into one item. Therefore, it is necessary to raise practicality by utilizing existing usability patterns [12], which is a future work.

5 Related Work

Boronat and colleagues [13] have proposed integration method of a class diagram as model generation support and Mehra and colleagues [14] have proposed cooperation and synthesis model creation support method for differences of plural figures. Through these methods, it is expected that inconsistencies between models can be resolved. On the other hand, our tool generates a STM from a conceptual model, and it is possible to match between conceptual models and screen transition models. Moreover, since it is possible to omit the trouble of generating a STM by full scratch and the efficiency of model generation increases, the tool is considered to be effective as model generation support.

Deeptimahanti and colleagues [15] have proposed a method to semi-automatically generate UML models from natural language requirements. If UML model can be generated from natural language requirements, developers with a little development experience can describe in natural language, it is possible to easily generate models. However, the UML model generated from natural language requirements does not include models representing screen structures or transitions. On the other hand, we can generate a STM based on a conceptual model by our proposed tool, and generate a layout model and prototype system by using existing tools. Therefore, it is considered that a STM can be automatically generated from natural language requirements by utilizing the method of Deeptimahanti et al. for this tool, and I want to make it a future work.

Miao and colleagues [16] have proposed a method of converting from a platform-independent model of user interface to various mainstream platform-specific models. By this method, if an application runs different platforms, you can easily create a platform-specific user interface. Since this method is the return of user interface, our proposed tool generates a STM from a conceptual model that is not related to user interface. Hence, the proposed tool can efficiently generate a model of user interface, it is considered effective for developers support.

6 Conclusion

In this paper, we propose a tool to automatically generate a STM from OCRUD diagrams in order to support generation of a STM for developers. As a result of the evaluation, it was suggested that the proposed tool can generate a STM from a conceptual model as the base. Therefore, this tool can be expected to be useful as support for writing STMs. In addition, it was found that required model elements may not be written because they forgot to write some model elements when the developers manually write a STM. Therefore, it was suggested that support for creating STMs is necessary for developers.

6.1 Future Work

A STM that can be generated automatically by the tool is a basic STM based on a conceptual model. Therefore, even if a prototype system is generated from the STM, its practicality is considered to be low. Hence, it is necessary to prepare usability patterns [12] as extension functions, and the developers need to be able to generate a STM by merely incorporating the usability patterns into the STM. As a first step, we need to organize usability patterns, which is a future work.

In addition, the range supported by the tool is until generation of a STM. It has poor practicality since a STM has a low penetration rate. Therefore, it is necessary to strengthen the convertibility to commonly known WebML and DSML, and implementation as a tool becomes a future work.

Acknowledgments. This work was supported by JSPS KAKENHI Grant Numbers JP16H02804, JP15K15972.

References

1. Da, S., Alberto, R.: Model-driven engineering: a survey supported by the unified conceptual model. Comput. Lang. Syst. Struct. **43**, 139–155 (2015)
2. Siqueira, F.L., Silva, P.S.M.: Transforming an enterprise model into a use case model in business process systems. J. Syst. Softw. **96**, 152–171 (2014)
3. Achouri, A., Ayed, L. J. B.: A formal semantic for UML 2.0 activity diagram based on institution theory. arXiv preprint arXiv:1606.02311 (2016)
4. Kamimori, S., Ogata, S., Kaijiri, K.: Automatic method of generating a web prototype employing live interactive widget to validate functional usability requirements. In: 3rd International Conference on ACIT-CSI, pp. 8–13 (2015)
5. Yazawa, Y., Ogata, S., Okano, K., Kaiya, H., Washizaki, H.: Traceability link mining–focusing on usability. In: 2017 IEEE 41st Annual Computer Software and Applications Conference (COMPSAC), vol. 2, pp. 286–287 (2017)
6. Ma, Z., Yeh, C. Y., He, H., Chen, H.: A web based UML modeling tool with touch screens. In: Proceedings of the 29th ACM/IEEE International Conference on Automated Software Engineering, pp. 835–838 (2014)
7. Krahn, H., Rumpe, B., Völkel, S.: Roles in software development using domain specific modeling languages. arXiv preprint arXiv:1409.6618 (2014)
8. Change Vision: Astah community-free UML modeling tool. http://astah.change-vision.com/ja/
9. Kaur, H., Singh, P.: UML (Unified Modeling Language): standard language for software architecture development. In: International Symposium on Computing, Communication, and Control (2011)
10. Torim, A.: A visual model of the CRUD matrix. Inf. Model. Knowl. Bases XXIII **237**, 313–320 (2012)
11. Akase, T., Ogata, S., Okano, K.: A support method for designing GUI consistent with screen transition model. IEICE Techn. Rep. **115**(487), 131–136 (2016)
12. Roder, H.: Specifying usability features with patterns and templates. In: 2012 First International Workshop on Usability and Accessibility Focused Requirements Engineering (UsARE), pp. 6–11 (2012)
13. Boronat, A., Carsí, J.Á., Ramos, I., Letelier, P.: Formal model merging applied to class diagram integration. Electr. Notes Theoret. Comput. Sci. **166**, 5–26 (2007)
14. Mehra, A., Grundy, J., Hosking, J.: A generic approach to supporting diagram differencing and merging for collaborative design. In: Proceedings of the 20th IEEE/ACM International Conference on Automated Software Engineering, pp. 204–213 (2005)
15. Deeptimahanti, D.K., Sanyal, R., Deeptimahanti, D. K., Sanyal, R.: Semi-automatic generation of UML models from natural language requirements. In: Proceedings of the 4th India Software Engineering Conference, pp. 165–174 (2011)
16. Miao, G., Hongxing, L., Songyu, X., Juncai, L.: Research on user interface transformation method based on MDA. In: 2017 16th International Symposium on Distributed Computing and Applications to Business, Engineering and Science (DCABES), pp. 150–153 (2017)

3D Formation Control of Swarm Robots Using Mobile Agents

Hideaki Yajima[1], Tadashi Shoji[1(✉)], Ryotaro Oikawa[1], Munehiro Takimoto[1], and Yasushi Kambayashi[2]

[1] Department of Information Sciences, Tokyo University of Science, Noda, Japan
6318515@ed.tus.ac.jp, mune@rs.tus.ac.jp
[2] Department of Computer and Information Engineering, Nippon Institute of Technology, Miyashiro, Japan
yasushi@nit.ac.jp

Abstract. In this paper, we propose a control algorithm to compose a arbitrary three dimensional formations which consists of actual swarm robots. The swarm robots that are composing formations such as specific polyhedrons or spheres coordinate each other by using network communication. Our control algorithm achieves the network communication by mobile software agents. A mobile software agent introduces control programs to each robot that has no initial program. In our formation algorithm, a formation is achieved through moving of a mobile agent called Ant agent to its own location. The movement process results in formation of robots because an Ant agent has to drive a robot in order to reach some specific place. In the process, a mobile agent can exchange a robot to drive through migration to another robot, which enables the agent to move without interference from other robots. This movement property makes formation more efficient, and contributes to formation of shapes filled inside. Also, in our algorithm, each agent does not have to know the absolute coordination of its own location. Instead, it knows the relative coordinates of its neighbor Ant agents, and it attracts them to the coordinates using Pheromone agent. A Pheromone agent repeatedly migrates between robots while adjusting the target coordinate. Once it finds a specific Ant agent to attract, it guides the Ant agent to the target coordinate. The guidance manner attracts neighbors to relative coordinates each other, enabling a shape to be composed without absolute coordinates. Therefore we can compose any formation at any coordinates in a three dimensional space. We have implemented a simulator based on our algorithm and conducted experiments to demonstrate the practical feasibility of our approach.

1 Introduction

Recent remarkable progress of hardware such as miniaturization and high performance of computer devices has made robot systems rapidly advanced not only the behavior of the robot but also the method of controlling the robot. One of such advanced control methods is an employment of mobile software

M. Virvou et al. (Eds.): JCKBSE 2018, SIST 108, pp. 168–177, 2019.
https://doi.org/10.1007/978-3-319-97679-2_17

agents, where an autonomous entity called agent controls physical mobile robot. In particular, the robot control systems based on multi-agents have introduced modularity, reconfigurability and extensibility for multi-robot systems [1]. Additionally, mobile software agents make robot control systems simple in distributed environments.

On the other hand, multi-agent systems may cause excessive interactions among agents, which require a rich network environment with wide bandwidth. Heavy communication load makes the multi-agent system less useful for controlling multiple robots. In order to mitigate the problems of excessive communication, mobile agent methodologies have been developed for distributed environments. In mobile agent system, each agent can move autonomously from a computer to another computer. Since a mobile agent can bring the necessary functionalities with it and performs its tasks autonomously, it can reduce the necessity for interaction between computers. In a case where communication is minimally performed, the mobile agent needs to establish a connection between computers only when it performs migration. It can be said that mobile agent system is particularly effective in ad-hoc network environment where connection can be intermittently established [2].

The migration property of the mobile agents contributes to the flexible and efficient use of the robot resources in multiple robots system. A mobile agent can migrate to the robot that is the most conveniently located for a given task. Since the agent migration is much easier than physical motion of a robot, the agent migration also contributes to saving power consumption [1]. In the environment where the tasks are coordinately executed by controlling a large number of robots with multiple agents, if the agents have a policy of choosing idle robots rather than busy ones, it would result in more efficient use of robot resources in addition to the power saving effect.

In this paper, we focus our attention on the formation control of mobile robots. It is one of the most important tasks that require collective actions. It is especially true where individual robot has limited abilities. For example, robots may aggregate for coordinated searching and rescue in the event of a disaster, collectively transporting a large object, pioneering and mapping unknown area, or maintaining a formation for the construction of buildings [3]. In most other formation control algorithms, each robot requires the information about the entire formation. In our algorithm, on the contrary, each robot requires only local neighbor information of the target formation.

We have pursued the idea of Ant Colony Optimization (ACO), and have implemented ants and pheromones in ACO as actual mobile software agents that control the mobile robots to compose a formation as in [4–6]. The ant agents attract each other using pheromone agents, and eventually settle down to appropriate locations composing the target formation. In our algorithm, the ant agents need only local information which consists of the relative positions to their adjacent robots. Each ant agent is instructed to drive a robot to the specific position. Therefore, it first looks for an idle robot that is located at the closest to the desired location through migrations. When it finds a convenient robot, it then starts driving to the position by using the given relative coordinates. Thus, the behaviors of mobile agents and robots driven by them enable a robot close to each part of a target formation to occupy the location for the part, contributing to efficient use of idle robots, efficient convergence, and suppressing energy consumption.

The reminder of this paper is organized as follows. In the second section, we describe the background. The third section describes method for composing a formation using mobile agents. The fourth section and the fifth section describe our mobile agent model and algorithm to implement in the real robots for formation control. In the sixth section, we report the results of numerical experiments using a simulator based on our algorithm under an environment constructed by data actually obtained through real equipment. Finally, we conclude our discussions in the seventh section.

2 Background

Takimoto et al. have presented a novel framework for controlling intelligent multiple robots [1,7]. The framework helps users to construct intelligent robot control software by migration of mobile agents. Kambayashi et al. also have presented a framework for controlling intelligent robots connected by the Internet [8]. The framework provides novel methods to control coordinated systems using higher-order mobile agents. Takimoto et al. and Nagata et al. have implemented a team of cooperative search robots to show the effectiveness of their framework [1,9]. In addition, they have demonstrated that their framework contributes to energy saving of multiple robots. Their approach has provided significant energy saving for search robot applications.

In recent years, an algorithm called ACO that is inspired by behaviors of social insects such as ants to communicate to each other by an indirect communication mediated by modifications of the environment called stigmergy is gradually established as a popular algorithm [10,11]. Goss et al. found that ants exchanged information by laying down a trail of a chemical substance called pheromone that is followed by other ants [12,13]. In order to solve the travel salesman problem, Dorigo et al. adopted an algorithm which was inspired by this ant's behavior today known as ACO called Ant Colony System (ACS) in the research [11]. This algorithm is used to solve various optimization problems, and many derivation algorithms are also proposed and considered as one of major research themes.

Oikawa et al. was inspired by ant's pheromone communication and proposed a multi-robots system composing formations [14]. In their method as well as Mizutani's proposal, they implemented ants and pheromones as mobile software agents and compose target formation. Ant agents generate pheromone agents that have local information about the formation to guide other ant agents to the appropriate locations. Once pheromone agents are generated, pheromone agents migrate to other robots to find the target ant agents. When ant agents receive the pheromone agents, the ant agents move robots to the locations indicated by the pheromone agents. They also proposed a control algorithm for composing formations of swarm robots based on their distributions using many mobile software agents [15]. In the approach, they introduced two kinds of mobile software agents as a guide agent and node agents. The guide agent traverses all the robots and calculates the conceptual barycenter of them, and then, calculates the suitable locations of the formation using the barycenter and generates node agents that drive robots to the calculated locations. Each node agent migrates to the robot that is nearest to the target location, and drives that robot to the location. Cheng et al. proposed a formation generation algorithm using Contained Gas Model in which robots act like a particle in a container [3].

Although these approaches have already been evaluated in numerical experiments by simulators, there are not many approaches that have been evaluated in practical experiments using actual machines or devices. Our approach assumes that the formation is given and the motions of robots are determined by pheromone. As well as formation control, we propose the techniques of generating a formation structure or determination of the individual location of robots in the formation.

3 Formation Using Mobile Agents

In our research, we assume that the robots can physically move to any coordinates in three dimension (3D), can measure distance and angle with an optical camera or other sensors, and can communicate each other through network such as WiFi. For example, we can envision image drones or autonomous underwater vehicles (AUVs) as the swarm robots. Our approach consists of two kinds of mobile software agents; namely ant agents (AAs) and pheromone agents (PAs). Each AA receives PAs, drives a robot along guidance of the PAs, and generates PAs to attract other AAs, which we call neighbor AAs, to occupy neighbor positions of the AA. On the other hand, each PA is generated by an AA. An AA always has a vector information pointing to a neighbor position of the AA. An AA traverses robots through Wi-Fi network while generating its own clones to search a neighbor AA to guide, and guides the neighbor AA based on its vector information, once it migrates to a robot that the neighbor AA drives. Local information is a set of vector values that represent the positional relationships between the neighbor AAs that compose the formation.

The process in which AAs compose the formation in cooperation with PAs is as follows. First, an AA traverses in a set of robots to find an idle robot in the field. Once it finds an idle robot, it occupies the robot to drive it, and generates PAs with vector value to each corresponding neighbor positions. Figure 1(a) shows each AA and its neighbor position as a gray circle and white circles respectively, and vector values from the AA to neighbor position as solid arrows. Second, PAs with the vector values repeatedly migrate to surrounding robots while generating their clones in order to reach the robot with the AA to guide. Figure 1(b) shows the situation where PAs have reached their target AAs. Red arrows show vector values that PAs hold. Finally, a robot with an AA settles to

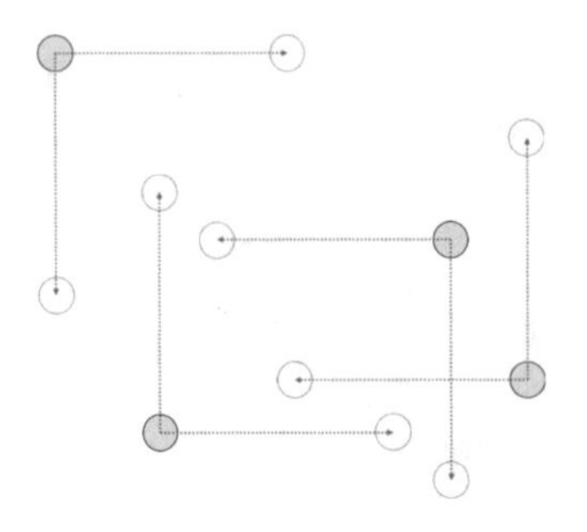

(a) Initial position of ant agents.

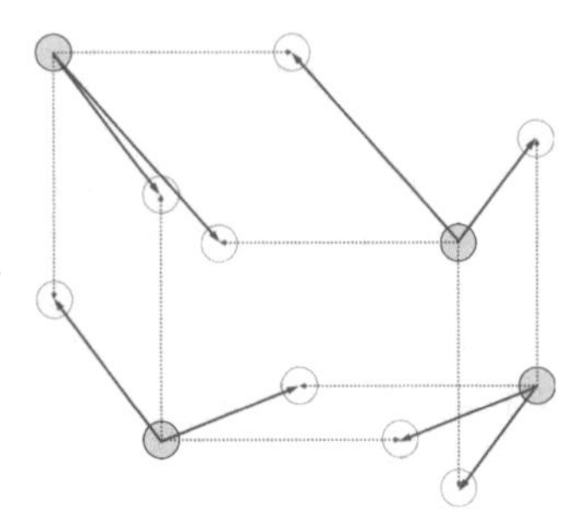

(b) Attracting target ant agents.

Fig. 1. Attraction through pheromone agents.

its own target position that satisfies the positional relationship with the neighbor robots, so that an objective square shape is composed.

4 Ant Agent

AAs occupy idle robots in the field and drive these robots along guidance of PAs they receive. AAs repeatedly migrate several robots while searching an idle robot through Wi-Fi network. Once the AA finds the idle robot, it occupies the robot, and generates PAs with vector values point to neighbor positions. After that, the PAs diffuse through migrations to find AAs to guide. On the other hand, an AA that receives the PA drives a robot to reach to the target position along guidance of the PA. If an AA receives n different source PAs with vector value V_i for the AA, these PAs are composed into a single PA with a vector value V_t as follows:

$$V_t = \frac{1}{n}\sum_{i=1}^{n} V_i \tag{1}$$

In the process where the AA approaches to its own target position along guidance of the PA, it does not drive a robot blindly. If the robot with the AA is far from the target position, the AA migrates another robot in the direction of the target instead of driving the current robot. This behavior contributes to the efficient motion to the target and suppressing energy consumption. However, in a situation where the robot is close to the target, the migration to another robot may cause the AA to pass through the target position. Thus, the AA judges whether to migrate to the other robot or drive the current robot by comparing V_t with vector value from another robot to the target position, V_{new}, which is calculated as follows:

$$V_{new} = V_t - V_{another} \tag{2}$$

As shown by Eq. 2, V_{new} can be calculated by V_t and a vector value from the current robot to another robot, $V_{another}$. If $|V_{new}|$ is less than half of the $|V_t|$, the AA migrates to the other robot; otherwise it drives the current robot along V_t.

5 Pheromone Agent

A PA is generated by an AA and has information of the neighbor locations of the AA and other neighbor AAs to occupy the desired location in the objective formation. The PA diffuses the information through migrations while generating its clones in the field to find the target AA to guide. First, the PA duplicates itself to make clones, which migrate to all the robots within the communication range of the current robot. Repeating the process, the PA reaches the target AA.

PAs are cloning themselves and diffusing their influences to other robots. As soon as PAs are generated by their source AA, they migrate to the robots in the communication range. If there is not the target AA on a robot where each PA migrates, the PA generate new clones, which further migrate to robots within the communication range. Because PAs are repeatedly generated in the diffusion process, PAs with the same source AA may reach a robot with the target AA through different routes. In such a case, the target AA updates the old PA to the last PA as a guide. PAs have duration time like physical pheromone, which gradually evaporates, so that all the PAs fade away over time. Thus, the information diffused by PAs is kept in fresh. The details of the information that a PA has are:

1. source ant agent ID,
2. target ant agent ID,
3. source ant agent angle value, and
4. vector value pointing to the neighbor position of the source AA, which the target AA should drive a robot to occupy.

The vector values of PAs must always point to the neighbor position, even if the PAs repeatedly migrate to other robots to find the target AA. Therefore, a vector value in a PA is updated to a new vector value $V_{new-target}$ whenever the PA migrates. The $V_{new-target}$ is calculated using V_t and a vector value $V_{another}$ from the current robot to another robot that is a destination of migration as follows:

$$V_{new-target} = V_t - V_{another} \tag{3}$$

6 Experimental Results

In order to prove the effectiveness of our approach in 3D environments, we have implemented our algorithm on a simulator and conducted numerical experiments. In the experiments, we assume the following conditions.

1. Robots are scattered in an $500 \times 500 \times 500$ cubic field in the simulator.
2. The range of Wi-Fi network for each robot is 150 units and the range of the sensor is the same units.
3. Each robot can move 2 units in each step in the simulator.

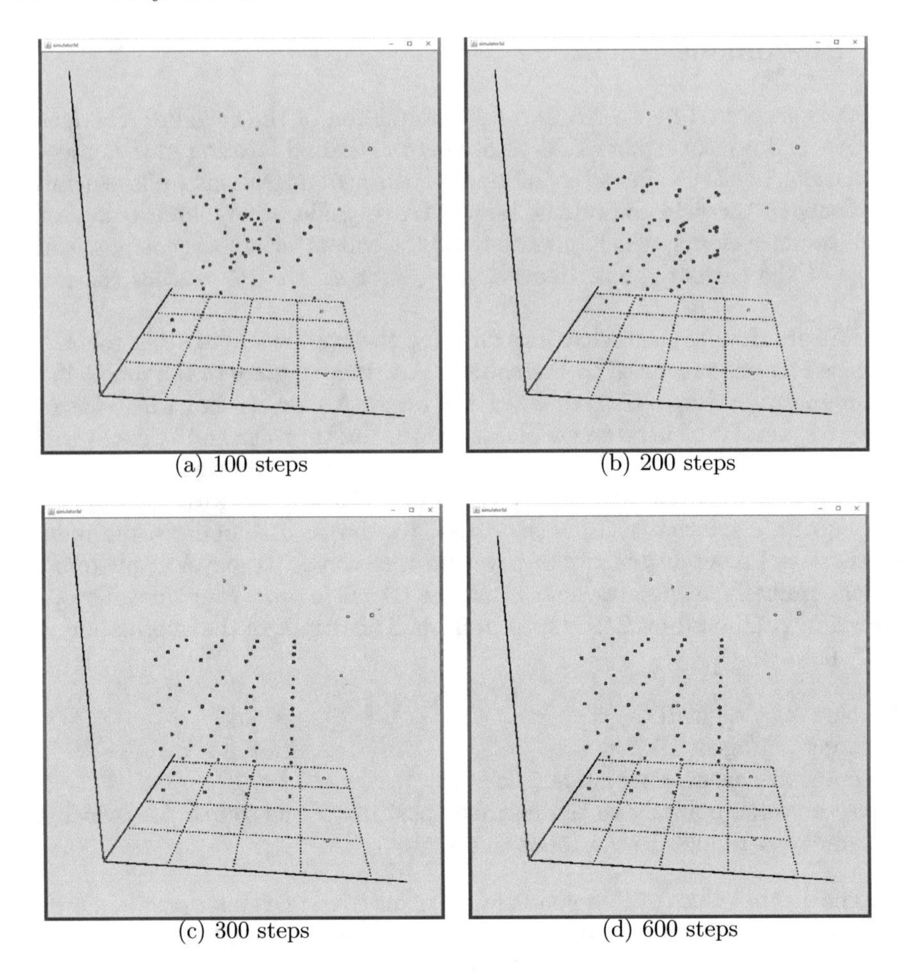

Fig. 2. Formation images for four kinds of steps

4. The initial locations and angles of robots are randomly selected without overlapping.
5. Each robot is represented as a circle on the grid field and emphasized as a red circle if it is driven by an AA.

We conducted experiments for a cube formation. Figure 2 shows images of simulation for a cube every 100 steps. Observing (a), (b) and (c) in this order, the shape of a cube is increasingly composed as step increases. Comparing (c) with (b), the shape at 300 steps is mostly same as the shape at 600 steps, so that we could say that this formation converges around 300 steps.

More quantitatively, we compared the completion rate of our formation. In our simulator, the formation task is completed when all the robots with an AA arrive at the target locations. In order to define the completion without ambiguity, we decide allowable deviation, and check whether the current locations of robots are under the deviation in every time steps. For checking the deviation, first, we calculated the barycenter of both actual and ideal positions, and then calculated all the coordinate of robots relative to the barycenter. Finally, we measured the distances between the relative coordinates of actual robots and the ideal relative coordinates, and calculated the averages $D_{average}$ of them for each shape with the different number of robots as follows:

$$D_{average} = \frac{1}{n}\sum_{i=1}^{n} \|I_i - A_i\| \tag{4}$$

In the equation, I_i is the ideal relative coordinate of a robot with an AA from the barycenter; A_i is the actual relative coordinate of the robot from the barycenter; and n is the number of AAs composing the formation.

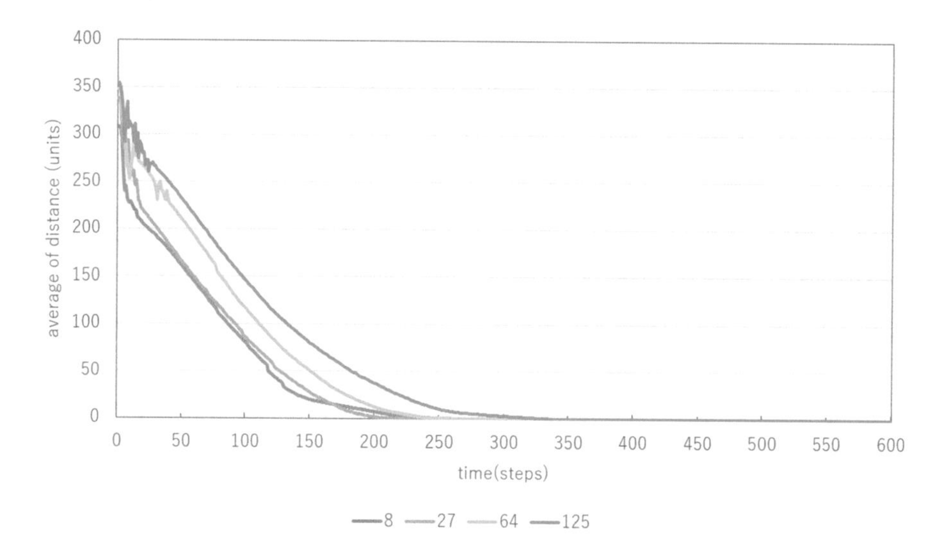

Fig. 3. Average of distances to target positions with different number of nodes.

Figure 3 shows $D_{average}$ for shapes with 8, 27, 64 and 125 nodes. In this simulation, it was 250 steps in the shape with 8 that is the minimum number of nodes, while it was 350 steps in the shape with 125 that is the maximum number of nodes. The rate of increase of completion time is not as much as that of the number of robots. The reason why our system can manage many robot is that we have integrated a clever way to manage robot collisions. The system manages a collision of robots by swapping the two driving agents, i.e. AAs so that it minimizes the physical movements of robots.

When we observe more details in Fig. 3, we can find that the average distances drastically decreases in the early stages, because the distances of AAs to the target positions are reduced through migrations. The decrease is particularly remarkable for the smaller number of nodes, such as 10 steps for the formation with 8 nodes, and 20 steps for the formation with 27 nodes, and it is also effective for composing formation with the large number of nodes.

7 Conclusion

We have proposed a three dimensional formation composing algorithm of real swarm robots using mobile agents. In the algorithm, we have implemented two kinds of mobile agents, ant agents and pheromone agents. Ant agents generate pheromone agents with local information that specifies where neighbor robots should occupy in the objective formation. Pheromone agents, which are generated by an ant agent, migrate to other robots in order to attract ant agents. Once the pheromone agent arrives in an ant agent, it specifies the ant agent's new position, and the ant agent starts driving the robot to the specified position to occupy. Eventually, all the robots occupy the right positions and compose the target formation. In order to demonstrate the effectiveness of our method, we have implemented a simulator and conducted numerical experiments. We have shown that our method can compose large formations without decreasing performance. We believe that our algorithm gives a small but an important contribution to the ubiquitous computing.

Acknowledgment. This work is partially supported by Japan Society for Promotion of Science (JSPS), with the basic research program (C) (No.17k01304), Grant-in-Aid for Scientific Research (KAKENHI).

References

1. Takimoto, M., Mizuno, M., Kurio, M., Kambayashi, Y.: Saving energy consumption of multi-robots using higher-order mobile agents. In: Proceedings of the First KES International Symposium on Agent and Multi-Agent Systems (KES-AMSTA 2007). Lecture Notes in Artificial Intelligence, vol. 4496, pp. 549–558. Springer, Heidelberg (2007)
2. Binder, W., Hulaas, J.G., Villazon, A.: Portable resource control in the J-SEA12 mobile agent system. In: Proceedings of the Fifth International Conference on Autonomous Agents (AGENTS 2001), pp. 222–223. ACM (2001)
3. Cheng, J., Cheng, W., Nagpal, R.: Robust and self-repairing formation control for swarms of mobile agents. In: Proceedings of the 20th National Conference on Artificial Intelligence - Volume 1 ser. AAAIf 05, pp. 59–64. AAAI Press (2005)
4. Mizutani, M., Takimoto, M., Kambayashi, Y.: Ant colony clustering using mobile agents as ants and pheromone. In: Proceedings of the Second Asian Conference on Intelligent Information and Database Systems Applications of Intelligent Systems (ACIDS 2010). Lecture Notes in Computer Science, vol. 5990, pp. 435–444. Springer, Heidelberg (2010)
5. Shintani, M., Lee, S., Takimoto, M., Kambayashi, Y.: Synthesizing pheromone agents for serialization in the distributed ant colony clustering. In: ECTA and FCTA 2011 - Proceedings of the International Conference on Evolutionary Computation Theory and Applications and the Proceedings of the International Conference on Fuzzy Computation Theory and Applications, Parts of the International Joint Conference on Computational Intelligence (IJCCI 2011), pp. 220–226. SciTePress (2011)

6. Shintani, M., Lee, S., Takimoto, M., Kamyabashi, Y.: A serialization algorithm for mobile robots using mobile agents with distributed ant colony clustering. In: Knowledge-Based and Intelligent Information and Engineering Systems. Lecture Notes in Computer Science, vol. 6881, pp. 260–270. Springer, Heidelberg (2011)
7. Kambayashi, Y., Ugajin, M., Sato, O., Tsujimura, Y., Yamachi, H., Takimoto, M., Yamamoto, H.: Integrating ant colony clustering to a multi-robot system using mobile agents. Ind. Eng. Manag. Syst. **8**(3), 181–193 (2009)
8. Kambayashi, Y., Takimoto, M.: Higher-order mobile agents for controlling intelligent robots. Int. J. Intell. Inf. Technol. (IJIIT) **1**(2), 28–42 (2005)
9. Nagata, T., Takimoto, M., Kambayashi, Y.: Suppressing the total costs of executing tasks using mobile agents. In: 42nd Hawaii International Conference on System Sciences 2009, HICSS 2009, pp. 1–10, January 2009
10. Dorigo, M., Birattari, M., Stützle, T.: Ant colony optimization-artificial ants as a computational intelligence technique. IEEE Comput. Intell. Mag. **1**(4), 28–39 (2006)
11. Dorigo, M., Gambardella, L.M.: Ant colony system: a cooperative learning approach to the traveling salesman. IEEE Trans. Evol. Comput. **1**(1), 53–66 (1996)
12. Goss, S., Aron, S., Deneubourg, J.L., Pasteels, J.M.: Self-organized shortcuts in the Argentine ant. Naturwissenschaften **76**, 579–581 (1989)
13. Goss, S., Beckers, R., Deneubourg, J.L., Aron, S., Pasteels, J.M.: How trail laying and trail following can solve foraging problems for ant colonies. In: Behavioural Mechanisms of Food Selection Volume 20 of the series NATO ASI Series, pp. 661–678 (1990)
14. Oikawa, R., Takimoto, M., Kambayashi, Y.: Distributed formation control for swarm robots using mobile agents. In: Proceedings of the Tenth Jubilee IEEE International Symposium on Applied Computational Intelligence and Informatics, pp. 111–116 (2015)
15. Oikawa, R., Takimoto, M., Kambayashi, Y.: Composing swarm robot formations based on their distributions using mobile agents. In: Multi-Agent Systems and Agreement Technologies, EUMAS: Lecture Notes in Computer Science, vol. 9571. Springer, Cham (2016)

Towards a Secure Semantic Knowledge of Healthcare Data Through Structural Ontological Transformations

Athanasios Kiourtis(✉), Argyro Mavrogiorgou, and Dimosthenis Kyriazis

Department of Digital Systems, University of Piraeus, Piraeus, Greece
{kiourtis,margy,dimos}@unipi.gr

Abstract. Current devices and sensors have revolutionized our daily lives, with the healthcare domain exploring and adapting new technologies, promising high quality of care. The diversity of healthcare data is leading to the independent operation of the latter, whilst the value emerging from their exploitation is limited. Most of the data is confined in data silos, without meeting the requirements of standards, and secure data exchange. Healthcare systems need to be able to communicate and exchange data anonymously, for better understanding about the results of prevention strategies, diseases, and efficiency of patient pathway management. Since healthcare interoperability is the only sustainable way to address these constraints, several techniques have been developed but they are partially applicable, rising the needs of a generic solution. The increased use of Electronic Health Records (EHRs) requires ontologies to capture domain knowledge, providing the basis for agreement within the healthcare domain. This manuscript focuses on the semantic interoperability of multiple EHRs – and their standards, proposing a way for primarily anonymizing the personal data stored into the EHRs, and then transforming the EHR datasets into XML Syntactic Models, for getting their semantic knowledge through their ontological representation.

Keywords: Interoperability · Heterogeneity · Anonymization · Ontologies
Electronic health records · Healthcare

1 Introduction

In recent years, discoveries and innovations in the healthcare coupled with the huge potential of information and communication technologies (ICT) have made personalized healthcare a possibility, accompanied with the creation of a variety of sensors and applications for supporting it [1]. Currently, the worldwide wearables market continues its upward trajectory with total shipment volumes reaching new records. According to data from the IDC Worldwide Quarterly Wearable Device Tracker, total volumes for the quarter reached 37.9 million units, up 7.7% from the 35.2 million units shipped in the same quarter a year ago [2]. Moreover, according to [3], the wearable devices market has generated a revenue of $30.5 billion in 2017.

M. Virvou et al. (Eds.): JCKBSE 2018, SIST 108, pp. 178–188, 2019.
https://doi.org/10.1007/978-3-319-97679-2_18

Currently, the data provided is heterogeneous and operate independently, whilst the value emerging from their exploitation is limited. Most of the traditional techniques are not able to deal with both the scale and the heterogeneity of wearable data in healthcare [4], thus it is getting common for preliminary indications of diseases to be missed. According to [5], medical errors in hospitals and other healthcare facilities are incredibly common and may now be the third-leading cause of death in the United States. A recent study [6] estimated that savings of approximately $78 billion could be achieved annually if data exchange standards were utilized across the healthcare sector. It should be mentioned that in 2013, $2.9 trillion were spent on healthcare, while due to the usage of data exchange standards, the upcoming years it led to between $348 billion and $493 billion in cost reductions [7].

Moreover, anonymizing health data is a challenging task due to the inherent heterogeneity, and can be considered as a critical piece of the healthcare puzzle, permitting the sharing of data for secondary purposes, mainly for data analysis and research. De-identification techniques are often at the forefront of companies' concerns when it comes to the processing of data. In addition, anonymization and pseudo-anonymization techniques have been a heavily debated topic in the ongoing reform of EU data protection law, thus the question that arises is how to implement these techniques in such a way that they will protect individual privacy, but still ensuring that the data is of sufficient quality [8].

The promise of a global standard for EHR records is still years away [9], as medical information systems need to be able to exchange complex and detailed medical data securely and efficiently. This can be achieved by constructing medical domain ontologies for representing medical terminology systems. The use of ontologies for representing biomedical knowledge is not new, since they have been widely used in biomedical domains for the last few years for different purposes [10, 11].

This manuscript addresses the aforementioned challenges by proposing a Structural Ontological Transformation process for transforming the collected health data into a common format through a five-step process, by identifying common links or similarities between different health data entities and HL7 FHIR resources [12]. Currently, this manuscript analyses the 1st, as well as the 2nd step of this Structural Ontological Transformation process. More particularly, the 1st step of the data pre-processing refers mainly to the anonymization of the healthcare dataset, prior to the 2nd step of its ontological transformation, in order to protect individual privacy, and make sure that the data is of sufficient quality. In that case, k-Anonymity [13] is implemented for impeding re-identification by removing some information, but letting the data to be intact for future use. With regards to the 2nd step, a formal syntactic and semantic model is represented for different EHR datasets of medical standards, where these datasets are expressed into ontological rules, including detailed and domain specific definitions of clinical concepts. Hence, the proposed approach facilitates the aggregation of the distributed heterogeneous data coming from multiple EHRs, so as to produce better-informed health related decisions and create new valuable information, supporting data sharing and integration for better decision making.

The rest of this manuscript is organized as follows. Section 2 describes the proposed approach for getting the ontological representation of different EHRs in combination with the pre-processing data anonymization step. Afterwards, a use case study following this approach is being provided in Sect. 3, while Sect. 4 addresses our challenges, analyzing our conclusions and plans.

2 Overall Architecture

The overall idea of the Structural Ontological Transformation process is to provide an automated way that is able to transform the structure of the ingested datasets into HL7 FHIR format, whereas through this implementation it will be possible for upcoming data to be automatically coordinated and distributed to the corresponding HL7 FHIR resources. In order to achieve the latter, five (5) different layers have been developed, each one responsible for providing different types of services (Fig. 1).

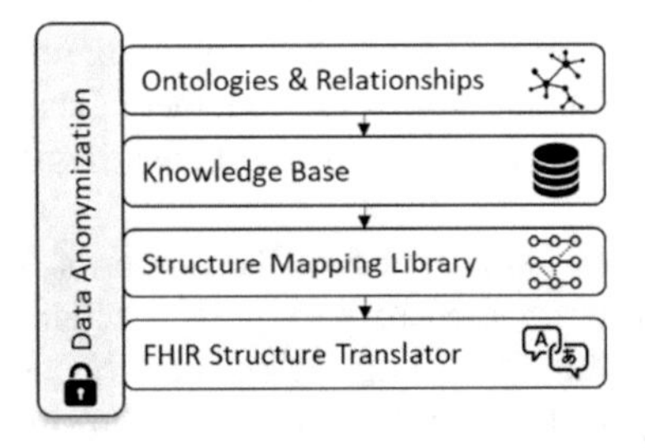

Fig. 1. Layers of the Structural Ontological Transformation process.

Prior to any transformation operation, the proposed approach involves the pre-processing of the healthcare data through the data anonymization mechanism, respecting privacy and security constraints. Shortly, the personal data that identifies the owner of the dataset is anonymized through k-Anonymity using the methods of data suppression and data generalization in the corresponding types of data. Since the data has been anonymized, the proposed approach continues by running ontology transformation operations in different health related datasets, storing the results into a Knowledge Base, in the form of triples of information. The results are then filtered and their semantics are identified, in order to discover the correspondence between the ingested data and a HL7 FHIR resource. To achieve that, probabilities of resemblance to a HL7 FHIR resource are assigned to each different element of the source data, while ontology alignment processes are triggered for the un-identified data elements. As a result, a direct mapping is made between the data elements in the source dataset and the HL7 FHIR resources, by implementing matching operations that run either in series or in parallel, in order to discover the semantics and/or the nature of the data elements. Queries are performed for accessing the Knowledge Base and retrieving the appropriate mapping and transformation information, to finally provide them as a response to the query and use it to transform the source dataset to the desired HL7 FHIR structure.

In the current manuscript, the pre-processing step of data anonymization will be thoroughly described, in which the healthcare data is anonymized. Accordingly, the Structural Ontological Transformation step will be also explained in details and will be performed as soon as the data has been fully anonymized, in order to respect and address the security and privacy constraints, which always constitute challenging tasks, and need to be considered when transmitting data. The overall mechanism, as well as the steps of the Knowledge Base, the Structure Mapping Library and the FHIR Structure Translator have been fully covered in one of our previous researches in [14].

2.1 Data Anonymization Layer

In the Data Anonymization layer, the data is pre-processed through k-Anonymity using the methods of data suppression and data generalization. In more details:

- Through the data suppression method, certain values of the attributes are replaced by a hashtag '#', according to their semantics and to what they represent. This method is usually implemented in the cases of textual data. For example, the value 'George' of the attribute 'Name' may be replaced by '#'.
- Through the data generalization method, individual values of attributes - mainly numeric values - are replaced, with a broader category, being given a range where the anonymized value can be found in between. For example, the value '23' of the attribute 'Age' may be replaced by '20 < Age < 30'.

Figure 2 represents the functionalities that are offered through the anonymization mechanism, consisted of the following steps:

1. Each personalized (i.e. de-anonymized) EHR dataset of each different standard is being processed through the *Data-type Identification* mechanism. The latter identifies whether an individual value of attribute can be anonymized through the data suppression or the data generalization method, by identifying the data type of each value. It should be mentioned that only the personal data (i.e. data that identify a person) are being filtered through this mechanism.
2. *k-Anonymity* is being implemented, where data generalization and/or data suppression are being applied, depending on the results of the *Data-type Identification* mechanism, resulting to the anonymized EHR dataset that will be provided as an input to the Structural Ontological Transformation process. The rule that this mechanism obeys to is that the numeric values are being anonymized through the data generalization method, while all the other types of values are being anonymized through the data suppression method.

2.2 Ontologies and Relationships Layer

In the Ontologies & Relationships layer, a comprehensive generic syntactic and semantic mechanism to cope with heterogeneous medical data deriving from different EHRs by obtaining the ontological representation of EHRs' data [15] is proposed, combining a series of technologies, namely Ontologies and XML Models. The input to this layer is an anonymized EHR dataset and the output is its ontological representation.

In more details, the developed mechanism (Fig. 3) provides a formal "Syntactic to Semantic" model for representing the EHR datasets of medical standards, through OWL Ontologies. With such a mechanism, the different EHR datasets are expressed into ontological rules, including detailed and Domain Specific definitions of clinical concepts. Creating the ontologies will be the same as deriving the relationships, classes and instances, and devolving this to triples of information. The main objective of this layer is to transform the data currently provided into an ontological form. Given that most of the data is represented through an XML/JSON format, it is difficult to create and visualize a hierarchical tree of the latter, in order to obtain a complete view of their relationships and inter-relationships. For that reason, the Ontologies & Relationships layer provides a way with which it will become feasible to identify the different relationships and rules that exist across the different data instances. Thus, the different relationships, classes and instances are discovered, providing a way for easily classifying the aforementioned into categories (i.e. families), and thus enabling easier identification and manipulation from the other existing layers. Based on the aforementioned, the Ontologies & Relationships layer consists of the following components:

1. *Anonymized EHR Data*: The Anonymized EHR Data component is assumed that it contains anonymized datasets of different medical standards that will be translated into OWL Ontologies through their transformation.
2. *XML Layer*: The XML Layer is based on the idea of Model-driven Engineering (MDE) [16] about using models at different abstraction levels for developing systems. In the current mechanism, model transformations expressed in XML will be used for the transformation of the EHR data to XML Models, while the latter will be represented into OWL Ontologies. In our case, an XML Translator is developed for transforming and translating the multiple types of EHR data into XML Models, which parses through the whole EHR dataset, converting the parent and child elements into XML format.

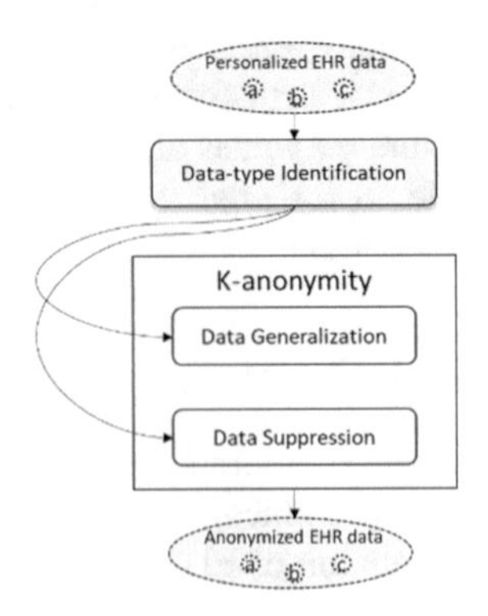

Fig. 2. Data Anonymization layer.

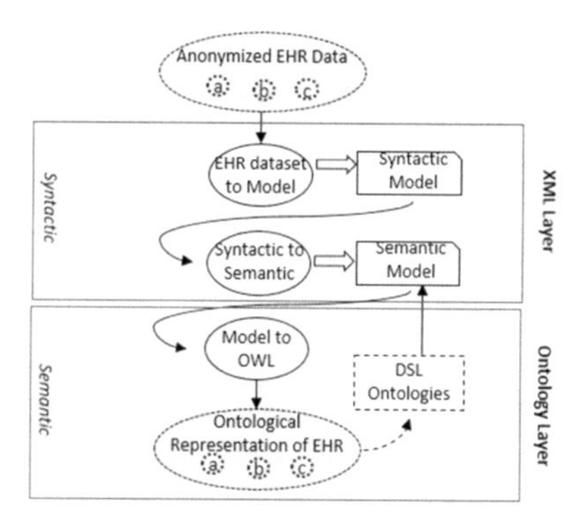

Fig. 3. Ontologies & Relationships Layer.

Having obtained the XML representation of the EHR data, the next step of the developed mechanism is to gather and store the nodes that constitute the current XML Model, through a Node Parser that finally generates the XML Nodes Model. In short, the Node Parser parses through the previously created XML Model, identifies and stores the parent and the child nodes, by bypassing the contents, the attributes and the data of these nodes. At this step, all of the created XML Models are given in their Syntactic expression to the Ontology Layer.

3. *Ontology Layer*: The Ontology Layer provides the formal semantics of our domain, and is composed of a series of OWL Ontologies developed for the multiple EHR medical standards. In more details, the Ontology Layer is built by identifying the common and disjoint knowledge defined in the ontologies of the multiple EHR medical standards, so it could be considered as a global EHR ontology. In the Ontology Layer, the detection of the equivalent concepts and data types is supported by the ontology integration methodology developed in [17]. In our case, the XML Nodes Models that are constructed into the XML Layer are being transformed into the ontological Class Hierarchy Model, following the work of [18]. In short, the developed algorithm parses the XML Nodes Model, and initializes the environment by importing the namespaces' prefixes to the ontological model that will be created. Furthermore, it creates the ontological representation of the XML Nodes Model, providing the Class Hierarchy Model of the EHR datasets through the integration of OWL API - version 3.4.2 [19].
4. *Ontological Representation* of EHR: In the OWL EHR component, the datasets of the different medical standards have been transformed into their ontological representation, while they provide metadata information in order to form the DSL Ontologies.

As a result of the described process, the ontological representation of different medical standards has been formed, and can be integrated into different healthcare systems, so that heterogeneous EHR datasets can be transformed into understandable, interoperable and shareable knowledge. The next steps of the Structural Ontological Transformation process include the storage of the ontologies into a Knowledge Base, resulting into the construction of a FHIR-based dataset, based on the fusion of the results of the Structure Mapping Library and the FHIR Structure Translator.

3 Results

3.1 Dataset Description

The dataset used for this experiment (i.e. Use Case Dataset) consists of a selected instance derived from a Personal Health Record (PHR), provided by BioAssist [20]. The dataset (Fig. 4a) contains attributes about specific measurements of a patient, the dates measured, the responsible for the measurement, etc. The file format of the used dataset is JavaScript Object Notation (JSON).


```
{
  "resourceType": "Observation",
  "id": "5d0b1f0f-a6d9-4c13-86aa-ec00fbe0f5e4",
  "status": "final",
  "code": {
    "coding": [
      {
        "system": "http://bioassist.gr",
        "code": "5502",
        "display": "Glucose"
      }
    ],
    "text": "Glucose"
  },
  "effectiveDateTime": "2017-06-29T21:00:00.000Z",
  "subject": {
    "reference": "111288",
    "display": "John Ranias"
  },
  "performer": {
    "reference": "000526",
    "display": "James Techaky"
  },
  "valueQuantity": {
    "value": "139.00",
    "unit": "mg/dl"
  },
  "referenceRange": [
    {
      "low": "70.0",
      "high": "110.0",
      "type": "normal",
      "text": "70.0 - 110.0 mg/dl"
    }
  ],
  "interpretation": "out of range"
}
```


(a)


```
{
  "resourceType": "Observation",
  "id": "5d0b1f0f-a6d9-4c13-86aa-ec00fbe0f5e4",
  "status": "final",
  "code": {
    "coding": [
      {
        "system": "http://bioassist.gr",
        "code": "5502",
        "display": "Glucose"
      }
    ],
    "text": "Glucose"
  },
  "effectiveDateTime": "2017-06-29T21:00:00.000Z",
  "subject": {
    "reference": "000001-222222",
    "display": "###"
  },
  "performer": {
    "reference": "000526",
    "display": "James Techaky"
  },
  "valueQuantity": {
    "value": "139.00",
    "unit": "mg/dl"
  },
  "referenceRange": [
    {
      "low": "70.0",
      "high": "110.0",
      "type": "normal",
      "text": "70.0 - 110.0 mg/dl"
    }
  ],
  "interpretation": "out of range"
}
```


(b)

Fig. 4. (a) Use Case Dataset, (b) Anonymized Use Case Dataset

3.2 Data Anonymization Results

The 1st step of the mechanism refers to the pre-processing of the Use Case Dataset, and more specifically to the anonymization of the patient's personal data, following the techniques described. The results of the anonymization are depicted in Fig. 4b.

3.3 Ontological Transformation Results

The 2nd step of the overall mechanism is to transform the anonymized Use Case Dataset into an XML Model, through the XML Translator. The result of the transformation is depicted in Fig. 5a. The 3rd step of the mechanism has to do with the transformation of the XML Model into the XML Nodes Model, through the Node Parser. The result of the transformation is depicted in Fig. 5b. The 4th step copes with the transformation of the

XML Nodes Model to the ontological Class Hierarchy Model. A snapshot of the transformation's result is depicted in Fig. 5c. Figure 5d represents finally, the visualization of the anonymized Use Case Dataset, based on its ontological Class Hierarchy Model, using Protégé OntoGraf [21], a plugin that gives support for interactively navigating the relationships of OWL ontologies.

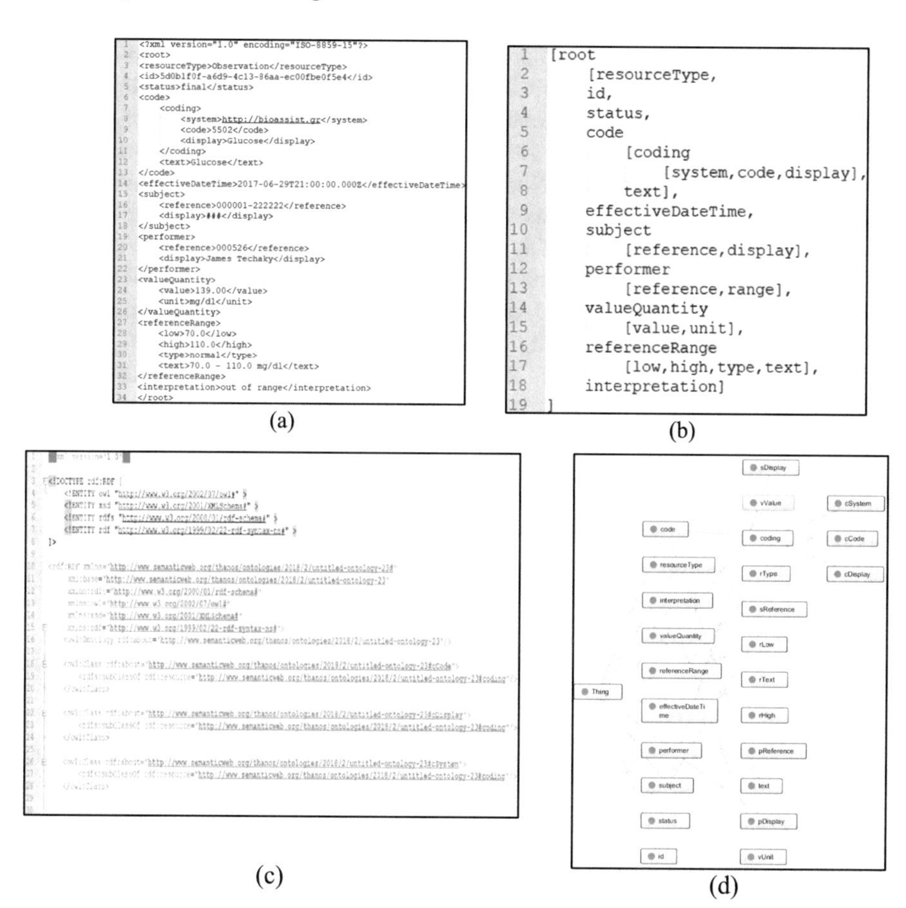

Fig. 5. (a) XML Model, (b) XML Nodes Model, (c) Snapshot of the Ontological Class Hierarchy Model, (d) Ontological Visualization.

4 Conclusions and Discussion

Currently, several solutions and researches have been conducted to help and provide access to the patient clinical information for different clinical organizations, towards quicker and more efficient decision-making. Among the basic disadvantages of these approaches is that they do not propose solutions easily applied to different medical standards and incoming data. Reference [22] is an example of research efforts that deal with integrating medical standards in order to ensure the interoperability between the

HL7 and the IEEE 1451 standards. The authors in [23] proposed an ontology to describe the patient's vital signs and to enable semantic interoperability when monitoring patient data. In the same direction, [24] proposed an ontology-driven interactive healthcare with wearable sensors, for acquiring context information at real time using ontological methods. Therefore, it becomes clear that there emerges a great need for creating a generic approach that can address and aggregate multiple groups of heterogeneous EHR data, to gain knowledge and offer a greater value to the patients' health ecosystem.

In this manuscript, in order to address and support this gap and define standard concepts and rules for this data, a global approach was proposed for identifying common links or similarities between different health data entities and as a result between their corresponding datasets. Specific techniques for the interoperable use of data in different services, locations, and contexts have been implemented, combined with an ontology transformation approach for achieving semantic interoperability. In our approach, a pre-processing step for anonymizing the personal data that identifies a patient through k-Anonymity has been implemented. Afterwards, the primary step of a complex process [14] has been thoroughly explained, following the research conducted in [15, 25, 26]. In short, the overall process employs several matching operations in the ingested healthcare data, and then filters the results of these matchers to find an overall alignment between them and the HL7 FHIR resources. The outcome is reflected in the restructuring of the ingested data, and the provision of a direct mapping between the latter and the globally used HL7 FHIR format. In the primary step that is explained in this manuscript, after the pre-processing step, XML Models and ontologies were used to provide a formal "Syntactic to Semantic" model for representing EHR datasets of multiple medical standards. As a result, a translation mechanism for transforming heterogeneous medical data into a common ontological representation was built, using OWL.

Henceforth, the presented implementation is an innovative approach for achieving data interoperability between multiple medical standards, as recent researches are designed to give a solution only to specific problems. On top of that, the data anonymization part adds a big advantage in our approach as it combines both the interoperability and the security of the same data. However, it must be noted that the mappings and the transformations could have been defined between the XML Models rather than the Ontologies, but in that case the mappings would be conceptual, without being linked to a particular ontology model such as the one provided by OWL.

The main objective remains to share the same semantics of data to ensure a smart interpretation for better decisions. Currently, we are working on the evaluation of different mapping languages, since having a formal representation of these mappings would be useful for extending our approach to additional medical standards. Furthermore, we are evaluating the developed mechanism, by testing it with multiple EHR datasets. Our future work focuses on the definition of more elaborated scenarios by collaborating with medical experts. while we aim to develop a generic and extensible architecture, capable of dealing with additional medical standards, addressing security and privacy constraints, through additional anonymization techniques.

Acknowledgments. The CrowdHEALTH project has received funding from the European Union's Horizon 2020 research and innovation programme under grant agreement No. 727560.

References

1. Personalized Health Care. http://qnphc.org/personalized-health-care/. Accessed 04 May 2018
2. Global Wearables Market Grows 7.7% in 4Q17 and 10.3% in 2017 as Apple Seizes the Leader Position, Says IDC. https://www.idc.com/getdoc.jsp?containerId=prUS43598218. Accessed 04 May 2018
3. Wearables market to generate 30.5 billion dollars in revenue (2017). https://fashionunited.uk/news/business/wearables-market-to-generate-30-5-billion-dollars-in-revenue-in-2017/2017082725609. Accessed 04 May 2018
4. Kalra, M., Lal, N.: Data mining of data with research challenges. In: CDAN (2016)
5. Medical errors now 3rd leading cause of death. https://www.washingtonpost.com/news/to-your-health/wp/2016/05/03/researchers-medical-errors-now-third-leading-cause-of-death-in-united-states/?utm_term=.896b9e01207b. Accessed 04 May 2018
6. Mead, C.N.: Data interchange standards in healthcare IT-computable semantic interoperability. In: JHIM, pp. 71–78 (2006)
7. How Big Data can help save $400 billion in healthcare costs. https://www.cio.com/article/2993986/big-data/how-big-data-can-help-save-400-billion-in-healthcare-costs.html. Accessed 04 May 2018
8. El Emam, K., Arbuckle, L.: Anonymizing Health Data: case studies and methods to get you started, 2nd edn. O'Reilly Media, Inc: 1005 Gravenstein Highway North, Sebastopol, CA95472, United States of America (2013)
9. Big Data Will Make a Difference in Saving Lives. https://www.linkedin.com/pulse/big-data-make-difference-saving-lives-pete-ianace. Accessed 04 May 2018
10. del Carmen Legaz-García, M., Martínez-Costa, C., Menárguez-Tortosa, M., Fernández-Breis, J.T.: A semantic web based framework for the interoperability and exploitation of clinical models and EHR data. In: Knowledge-Based Systems, pp. 175–189 (2016)
11. How Big Data and Ontology will improve healthcare. https://www.nomagic.com/news/insights/how-big-data-and-ontology-will-improve-healthcare. Accessed 04 May 2018
12. HL7 FHIR. https://www.hl7.org/fhir/resourcelist.html. Accessed 04 May 2018
13. Sweeney, L.: k-anonymity: a model for protecting privacy. Int. J. Unc. Fuzz. Knowl. Based Syst. **10**(5), 557–570 (2002)
14. Kiourtis, A., Mavrogiorgou, A., Kyriazis, D.: Structurally Mapping Health Data to HL7 FHIR through Ontology Alignment (under review)
15. Kiourtis, A., Mavrogiorgou, A., Kyriazis, D.: Aggregating Heterogeneous Health Data Through an Ontological Common Health Language. In: DeSE (2017)
16. da Silva, A.R.: Model-driven engineering: A survey supported by the unified conceptual model. JCL **43**, 139–155 (2015)
17. Fernández-Breis, J.T., Martiinez-Bejar, R.: A cooperative framework for integrating ontologies. IJHC **56**(6), 665–720 (2002)
18. Ontmalizer. https://www.openhub.net/p/ontmalizer. Accessed 04 May 2018
19. OWLAPI. https://www.w3.org/2001/sw/wiki/OWLAPI. Accessed 04 May 2018
20. BioAssist. https://bioassist.gr/#/?lang=en. Accessed 04 May 2018
21. Ontograf. https://protegewiki.stanford.edu/wiki/OntoGraf_1.0.1. Accessed 04 May 2018
22. Kim, W., Lim, S., Ahn, J., Nah, J., Kim, N.: Integration of IEEE 1451 and HL7 exchanging information for patients' sensor data. J. Med. Sys. **34**(6), 1033–1041 (2010)
23. Lasierra, N., Alesanco, A., Guillen, S., Garcia, J.: A three stage ontology-driven solution to provide personalized care to chronic patients. JBI **46**(3), 516–529 (2013)
24. Kim, J.H., Lee, D.S., Chung, K.Y.: Context-aware based item recommendation for personalized service. In: ICISA, pp. 595–600 (2011)

25. Kiourtis, A., Mavrogiorgou, A., Kyriazis, D.: Gaining the semantic knowledge of healthcare data through syntactic models transformations. In: ISCSIC, pp. 102–107 (2017)
26. Kiourtis, A., Mavrogiorgou, A., Kyriazis, D.: Towards data interoperability: turning domain specific knowledge to agnostic across the data lifecycle. In: WAINA, pp. 109–114 (2016)

Group Affect Recognition: Optimization of Automatic Classification

Andreas M. Triantafyllou and George A. Tsihrintzis(✉)

Department of Informatics, University of Piraeus, Piraeus, Greece
aandreasam@gmail.com, geoatsi@unipi.gr

Abstract. In this paper, we present the next step in our research on detecting emotions of groups of people from a computer, with special emphasis on educational technologies. Specifically, we present our results towards optimization of the automated classification method. In previous experiments for collecting, exporting, and classifying faces to create databases of group sample, we conducted evaluation of the basic ways of automatic classification of faces. From the evaluation results, we concluded that additional approaches were needed towards optimization of the classification algorithms, as the error rates of the basic automated classification approaches were not negligible. We present the optimization approaches we considered and evaluated, which result in both lower error rates and decrease in the required time of creation of databases with group samples. As a result, the automatic processes of correct classification and quick creation of complete databases with group samples provide a significant contribution in the field of emotion detection of groups of people from computer.

1 Introduction

The idea for the research and implementation of new procedures in relation to the automatic classification of faces, came from the need to reduce the error rate of classification. By reducing the error rate, we manage to reduce the build time of completed databases with group samples.

First we briefly mention the basic automated classification process as we have seen in [2] Group affect Recognition: Visual - Facial Data Collection. Then mention evaluation as we saw in [3] Group affect Recognition: Evaluation of basic automated sorting, so that we get to the object of this paper and the continuation of research to optimize the automated classification method. After the brief presentation of the previous work, we present the optimization methods for automated classification as well as the presentation of the extensions of our software.

Continuing, we present the use of optimized methods in experiments data. Through the results of the use of the new methods, we analyze the statistics of the success of the classifications, compare them and finally present the overall evaluation. Having now available the software for faces classification, with the

This work has been partly supported by the University of Piraeus Research Center.

M. Virvou et al. (Eds.): JCKBSE 2018, SIST 108, pp. 189–196, 2019.
https://doi.org/10.1007/978-3-319-97679-2_19

lowest error rate, we can easily build large completed databases with group samples in much less time. The faster creation of completed databases with group samples will greatly assist us in the research to detect human group emotions from computers targeted at educational events.

2 Previous Related Work

In previous works, they have dealt extensively with the field of emotion detection with various approaches of a human from a computer. Other works by drawing conclusions from face images, others with meaning extraction from written texts etc.

We have started research to detect the emotions of a group of people from computers who are attending an educational event. For the approach to detecting human group emotions, we need to work on specific datasets. In particular, we need data from samples correlated with each other based on time, depth of time, and each educational event. This approach has been taken from our first paper [2] Group affect Recognition: Visual - Facial Data Collection, in which we have defined the group samples, we have collected data through experiments in real conditions and we have implemented procedures to help us, thus laying good groundwork in our research. Then we proceeded to the evaluation of basic automatic classification of faces in [3] Group affect Recognition: Evaluation of basic automated sorting. Automatic classification of faces helps us to create completed databases with group samples to be able to create emotion detection algorithms using the data for training.

Based on the evaluation of the basic automated classification method, we have come to the conclusion that we need to investigate solutions to optimize the process. In this work, we propose the solutions we have researched and implemented, managing to reduce the build time of completed databases with group samples.

3 Optimization Ideas and New Methods

To proceed with the presentation of new ideas and methods to reduce the error rate of automatic face classification, we will make a reference to the basic automated classification we have seen in [2] Group affect Recognition: Visual - Facial Collection. In basic automated classification we had an adjustable comparison threshold proportional to the current frame, the first detection of faces from each event and the changes in the crowd of people attend it. In this way, the software started without any detectable face and tried to detect them from the beginning by classifying the different faces according to the comparison thresholds.

As we have seen in the basic automated classification evaluation in [3] Evaluation of basic automated sorting, the error rate is around 0.21–0.25. As we understand this percentage is large enough to manually correct the wrongly sorted samples in large numbers of samples. For this reason, we had to investigate and implement ways to reduce the error rate so that the total creation time of completed databases with group samples is reduced enough. Our initial thought to improve the error rate of automatic classification was to help the software by giving it samples from the beginning, of the faces who attending the educational event. In this way, the software has a higher percentage of success because it is easier to match a face to a multitude of faces than not having them and needing to find them first. In the case of sample detection to create completed databases with group samples for training, it was real-time, then the students who attended the educational event, they should give samples of their faces before the beginning of the course so that to have the software at its disposal from start. In our case, the detection and classification of faces for the creation of completed databases is done through videos that we have recorded in various educational events. That is why we thought in the first frames to stop the program according to the sampling parameters and to do manual tagging at the first occurrences of the faces. Then we decided to combine this method with adjustable comparator thresholds, so that in large differences the program to make automated tagging.

The second method we thought and implemented is in fact an extension of the first. That is, instead of doing manual tagging on the first appearance of the faces (beyond the samples that the program does for automated tagging) and these being the set of training, to we do manual tagging and to some subsequent appearances. In this way we grow up the training set and we have more faces (with possible slightly different facial attitudes) for each student. Also, in this way we are able to increase the range of comparisons and reduce the likelihood of a classification error in different postures from the training set. And in this method we keep the adjustable thresholds and the ability of the program to do automated tagging.

After the presentation of methods in software extensions, we will use them to data we have collected from experiments, and then compare and evaluate the new methods.

4 Software Extensions

In this section of our paper, we will present the software extensions we have implemented for the new automated sorting methods that we have seen above.

We first created a new window to support the functionality of manual tagging. In this window there are an image of the person we want to tag, a textbox for the tag, a button for save, and a scrollviewer with other face images with their tags.

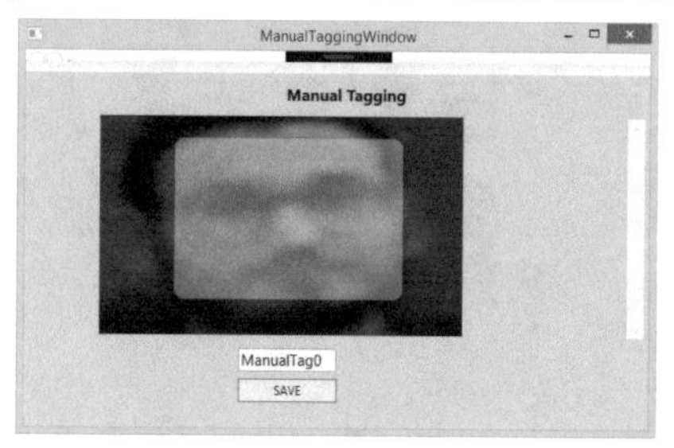

Window for the munal tagging for the first face. Has no other person tagged.

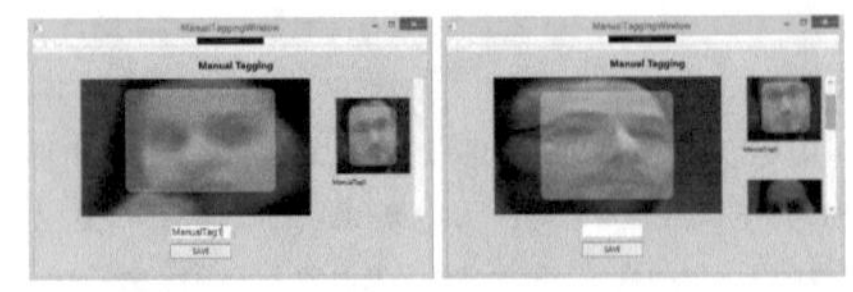

In the right column we see other faces with their tag, for the case that we see a face by another tag and we do not remember its label.

Also in relation to the new methods we have implemented the backend functionality and we have adapted the program flow according to our options.

Specifically for the first method we perform manual tagging by getting 1 sample for each face in the training set. Also operate the adjustable comparison thresholds for automatic tagging.

For the Second Method, we keep the adjustable comparison thresholds for automatic tagging, and we also manually tagging the faces' initial impressions, getting more than one image of each face for each tag in the training set.

5 Use of the New Methods

Moving on to this section of our work, we will present the use of our software for taking samples from lecture videos and classifying them based on the new methods we mentioned above.

Specifically on the screen with the data collection parameters, we have added a checkbox to select the second new method we have implemented. Whether its first version (with one sample for each face in the training set) or the second version (with more than one sample for each face in the training set).

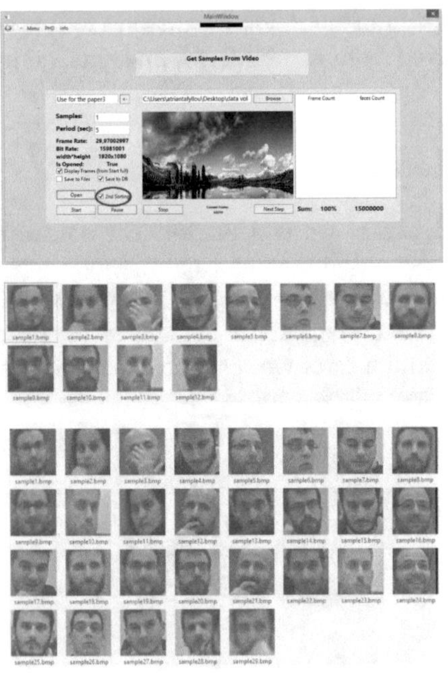

The training sets from the first and second methods respectively.

The rest of the program is used in the same way as we have seen in previous [1,2] papers. That is, we select the video from the educational event we want, we enter the parameters and then we start the process.

6 Evaluation of New Automated Sorting and Conclusions

After the presentation of the new methods of optimized automatic classification, we continue to evaluate them. For the purposes of the evaluation we use the functions we have created in [3] Group impact Recognition: Evaluation of basic automated sorting.

Specifically, we collect samples from common videos from educational events and perform the automatic classification in the new ways.

In the sample collection and classification with the new methods, we use common parameters. Then, from statistics extraction functions, we collect all the statistics and compare their values.

Indicative, some statistics from the new method 1 and the new method 2 respectively.

After several statistical collections and various tests we found that the error rates of the two new methods range in:

- Method1 ~ **0,14–0,18**
- Method2 ~ **0,05–0,09**

The second new method, i.e. the method that holds more than one sample for each faces in the training set, has a lower error rate and this method is kept for the continuation of our research. However, both new methods of automatic classification of faces are better than the basic automatic classification that we have seen in [2] Group affect Recognition: Visual - Facial Collection and it has error rate ~ 0.21–0.25 from [3] Group affect Recognition: Evaluation of basic automated sorting.

7 Synopsis and Future Work

Summarizing, in this paper we have presented the need born that from [3] Group Affect Recognition: Evaluation of basic automated sorting, to investigate and to implement new methods of automatic classification of faces in order to reduce the error rate of the basic automated classification way.

Initially, we briefly outlined the our previous works in relation to the emotions detection of group of people from a computer who attending educational events, as well as we presented and the basic automated classification method with the adjustable thresholds of comparison. The goal - the need for automatic classification of faces is to we can easily and quickly build databases with large numbers of group samples, which will be useful for developing algorithms for emotions detection of groups of people who attending educational events. From the basic automated classification method, the error rate for the classification of faces to depth of time as we saw in [3], makes us quite time consuming to manually sort the wrongly sorted faces. For this reason we investigated and implemented two new methods:

- The first method, in the process of extracting samples from video from educational events, asks us in the first impressions of faces to we do manual tagging. We also adapted the functionality with the adjustable thresholds of comparison. In this method we hold a sample for each face in the training set.
- The second method is essentially an extension of the first. Its main difference is that it holds more than one sample for each face in the training set. In this way it can more easily make matching the faces who have differences in facial attitude with their respective tags.

We then evaluated the two methods and ended up holding the second method (which holds more than one sample for each face in the training set) as it had a lower error rate.

With this work we manage to achieve the realization of exporting group samples and creating databases at a very good level that allows us to continue in the next steps of our research. Specifically, in our future work, we will create completed databases with group sampes ready for use, so that they can help us in researching and creating emotional detection algorithms of computer based to groups of people who are attending educational events. Also the completed databases with group samples will help us to research to create possible patterns of comparison.

References

1. Triantafyllou, A: Detecting human emotion/mood from a computer. M.Sc. thesis (2017)
2. Triantafyllou, A.M., Tsihrintzis, G.A.: Group affect recognition: visual - facial collection. In: IEEE 29th International Conference on Tools with Artificial Intelligence (ICTAI) (2017)
3. Triantafyllou, A.M., Tsihrintzis, G.A.: Group affect recognition: evaluation of basic automated sorting. In: IEEE 9th International Conference on Information, Intelligence, Systems and Applications (IISA) (2018)

4. Virvou, M., Tsihrintzis, G.A., Alepis, E., Stathopoulou, I.-O., Kabassi, K.: Toward affect recognition by audio-lingual and visual-facial modalities: empirical studies and multi-attribute decision making. Int. J. Artif. Intell. Tools **21**(2), 1240001-1–1240001-28 (2012)
5. Veenendaal, A., Daly, E., Jones, E., Gang, Z., Vartak, S., Patwardhan, R.S.: Group emotion detection using edge detection mesh analysis (2014)
6. Suslow, T., Junghanns, K., Arolt, V.: Detection of facial expressions of emotions in depression (2001)
7. Graesser, A., McDaniel, B., Chipman, P., Witherspoon, A., DMello, S., Gholson, B.: Detection of emotions during learning with AutoTutor (2006)
8. Stathopoulou, I.-O., Alepis, E., Tsihrintzis, G.A., Virvou, M.: On assisting a visual-facial affect recognition system with keyboard-stroke pattern information. Knowl. Based Syst. Elsevier **23**, 350–356 (2010)
9. Stathopoulou, I.-O., Tsihrintzis, G.A.: Appearance-based face detection with artificial neural networks. Intell. Decis. Technol. **5**(2), 101–111 (2011)
10. Stathopoulou, I.-O., Tsihrintzis, G.A.: Visual affect recognition. Front. Artif. Intell. Appl. **214**. IOS Press (2010)
11. Stathopoulou, I.-O., Alepis, E., Tsihrintzis, G.A., Virvou, M.: On assisting a visual-facial affect recognition system with keyboard-stroke pattern information. Knowl.-Based Syst. **23**(4), 350–356 (2010)
12. Alepis, E., Stathopoulou, I.-O., Virvou, M., Tsihrintzis, G.A., Kabassi, K.: Audio-lingual and visual-facial emotion recognition: towards a bi-modal interaction system. ICTAI **2**, 274–281 (2010)
13. Stathopoulou, I.-O., Alepis, E., Tsihrintzis, G.A., Virvou, M.: On assisting a visual-facial affect recognition system with keyboard-stroke pattern information. In: SGAI Conference, pp. 451–463 (2009)
14. Tsihrintzis, G.A., Virvou, M., Alepis, E., Stathopoulou, I.-O.: Towards improving visual-facial emotion recognition through use of complementary keyboard-stroke pattern information. In: ITNG, pp. 32–37 (2008)
15. Tsihrintzis, G.A., Virvou, M., Stathopoulou, I.-O., Alepis, E.: On improving visual-facial emotion recognition with audio-lingual and keyboard stroke pattern information. In: Web Intelligence, pp. 810–816 (2008)
16. Stathopoulou, I.-O., Tsihrintzis, G.A.: NEU-FACES: a neural network-based face image analysis system. ICANNGA **2**, 449–456 (2007)
17. Stathopoulou, I.-O., Tsihrintzis, G.A.: Towards automated inferencing of emotional state from face images. In: ICSOFT (PL/DPS/KE/MUSE), pp. 206–211 (2007)
18. Stathopoulou, I.-O., Tsihrintzis, G.A.: A neural network-based system for face detection in low quality web camera images. In: SIGMAP, pp. 53–58 (2007)
19. Alepis, E., Virvou, M., Kabassi, K.: Affective reasoning based on bi-modal interaction and user stereotypes. New Dir. Intell. Interact. Multimedia **142**, 523–532 (2008)
20. Matsumoto, D., Ekman, P.: Facial expression analysis. Scholarpedia **3**(5), 4237 (2008)
21. Cohn, J.F., Schmidt, K.L., Gross, R., Ekman, P.: Individual differences in facial expression: stability over time, relation to self-reported emotion, and ability to inform person identification. In: ICMI, pp. 491–498 (2002)
22. Donato, G., Bartlett, M.S., Hager, J.C., Ekman, P., Sejnowski, T.J.: Classifying facial actions. IEEE Trans. Pattern Anal. Mach. Intell. **21**(10), 974–989 (1999)
23. Bartlett, M.S., Donato, G., Movellan, J.R., Hager, J.C., Ekman, P., Sejnowski, T.J.: Image representations for facial expression coding. In: NIPS, pp. 886–892 (1999)

24. Bartlett, M.S., Viola, P.A., Sejnowski, T.J., Golomb, B.A., Larsen, J., Hager, J.C., Ekman, P.: Classifying facial action. In: NIPS, pp. 823–829 (1995)
25. Kapoor, A., Qi, Y.A., Picard, R.W.: Fully automatic upper facial action recognition. In: AMFG, pp. 195–202 (2003)
26. Kapoor, A., Picard, R.W.: Real-time, fully automatic upper facial feature tracking. In: FGR, pp. 10–15 (2002)
27. Picard, R.W.: Toward computers that recognize and respond to user emotion. IBM Syst. J. **39**(3&4), 705–719 (2000)

An Artificial Immune System-Based Approach for the Extraction of Learning Style Stereotypes

Dionisios N. Sotiropoulos[1], Efthimios Alepis[1], Katerina Kabassi[2], Maria K. Virvou[1], and George A. Tsihrintzis[1](✉)

[1] Department of Informatics, University of Piraeus, 80, M. Karaoli & A. Dimitriou St., 18534 Piraeus, Greece
{dsotirop,talepis,mvirvou,geotsi}@unipi.gr

[2] Department of Environmental Technology, Technological Educational Institute of Ionian Islands, 29100 Zakynthos, Panagoula, Greece
kkabassi@teiion.gr

Abstract. This paper presents an unsupervised computational mechanism which exhibits the ability to reveal the inherent group structure of learning patterns that pervade a given set of educational profiles. We rely on the construction of an Artificial Immune Network (AIN) of learning style exemplars by proposing a correlation-based distance metric. This choice is actually imposed by the categoric nature of the underlying data. Our work utilizes an original dataset which was derived during the conduction of an extended empirical study involving students of the Hellenic Open University. The educational profiles of the students were built by collecting their answers on a thoroughly designed questionnaire taking into account a wide range of personal characteristics and skills. The efficiency of the proposed approach was assessed in terms of cluster compactness. Specifically, we measured the average correlation deviation of the students' education profiles from the corresponding artificial memory antibodies that represent the acquired learning style stereotypes.

Keywords: Educational profiles · Learning stereotypes · Clustering Artificial immune systems

1 Introduction

Online learning is pervading higher education, compelling educators to confront existing assumptions of teaching and learning in higher education [7]. This is especially the case for open universities, which have students of different ages, background knowledge and skills. For such students that work while studying an online course, the online learning environments provide flexibility regarding time and place as well as a self-paced learning [3].

M. Virvou et al. (Eds.): JCKBSE 2018, SIST 108, pp. 197–206, 2019.
https://doi.org/10.1007/978-3-319-97679-2_20

Although these advantages seem rather important, e-learning environment are sometimes regarded inefficient compared to the traditional learning alternative [10]. A way for developing efficient e-learning environments is when they provide individualized learning. For a system to adapt to each learner, it should incorporate a learner model.

A learner model can be described as a combination of personality factors, behavioral factors and knowledge factors [8]. Furthermore, as a part of personality factors, learners have different learning styles. Rich [13] identified one model of a single, canonical user vs. a collection of models of individual users. As a result, an e-learning system may maintain an individual user model or some user models that represent classes of users. These classes are called stereotypes.

Among the user features considered in user models, learning styles may play an important role in individualizing and improving learning [1]. A learning style is defined as characterizing strengths and preferences during the learning process [4]. Each student has his/her unique way of processing new information during the learning process. However, users may be grouped according to their common characteristics taking into account the special feature of each learning style. Therefore, we aim at designing some stereotypes that correspond to the possible learning styles of the learners.

In view of the above we have conducted an empirical study among the students of the Hellenic Open University that aims at designing the educational models of the students and especially the stereotypes of users according to their learning style. For this purpose, we have selected the Felder-Silverman Learning Style Model (FSLSM) proposed by Felder and Silverman [4]. The particular model seems especially detailed and comprehensive. Furthermore, it has been suggested by researchers [11] that the FSLSM seems to be particularly appropriate for being applied in e-learning systems.

The primary purpose of this paper lies upon the extraction of fundamental learning style stereotypes from a given dataset of educational profiles. The original data were obtained from students of the Hellenic Open University during the conduction of a thorough empirical study. Each student was asked to provide his/hers answers to an extensive questionnaire covering a wide spectrum of personal characteristics and learning skills. Our aim was to reveal the intrinsic structure of the student-specific educational feature vectors by adopting the computational framework of Artificial Immune Systems (AIS). Specifically, we build an Artificial Immune Network (AIN) of learning style stereotypes in a multi-dimensional vector space defined by a correlation-based distance metric. The resulting learning style exemplars coincide with the set of artificial memory antibodies which are generated by the AIS-based clustering algorithm. Thus, each artificial memory antibody provides a mathematical representation for the common learning style characteristics that pervade a given group of students.

The remainder of the paper is organized as follows: Sect. 2 presents the related work in learning styles and refers to some studies that aimed at finding the relationship between the learning style and the e-learning outcome. In Sect. 3 we analyze the setting of the experiment conducted for the empirical study. In

Sect. 4 we provide the details concerning the formation of the utilized dataset and the employed standardization procedure. The core of our algorithmic approach is described in Sect. 5 were we also elaborate on the proposed correlation-based distance metric, whereas Sect. 6 is devoted on discussing the obtained experimental results. Finally, in Sect. 7, we discuss the conclusions drawn from this work and present some future work.

2 Related Work

Several reviews of learning styles and classification of student exist in the literature [2,12]. The two more widely used learning styles models are Kolb's model and the Felder-Silverman Learning Style Model (FSLSM).

According to the Kolb's model, learning follows a cyclic pattern that is based on four stages Concrete experience, Reflective Observation, Abstract Conceptualization, Active experimentation. In view of these cycles, learners can be categorized into one of four styles: Diverger (preference for feeling and watching), Assimilator (preference for thinking and watching), Converger (preference for thinking and doing) and Accomodator (preference for feeling and doing). Several researchers have applied Kolb's learning styles [14]. Most of the researchers applying the model to e-learning systems have concluded that learning styles are a key factor in the effectiveness of learning [9].

FSLSM, on the other hand, describes the learning style of learners in four different dimensions. These dimensions are described below:

- Sensing/Intuitive perception: Sensing learners learn best when given facts and procedures whereas intuitive learners prefer concepts and interpretations.
- Visual/Verbal Input: Visual learners get more information from visual material such as pictures, diagrams, graphs, etc. Verbal learners, on the other hand, prefer verbal material such as written and spoken words and mathematical formulas.
- Active/Reflective Processing: Learners learn either actively via experiments and collaboration with others or reflectively when they work by themselves and without experiments.
- Sequential/Global Understanding: Learners understand a concept either sequentially by following step by step or globally by taking a holistic view of the concept and then going into details.

One of the main advantages of FSLSM, is that it is followed by the Index of Learning Styles (ILS) [6]. This index is mainly a questionnaire that contains 44 questions divided in four groups. Each group of 11 questions is used for evaluating one of the four dimensions that FSLSM uses for describing the learner's learning style. The validity and reliability of the ILS were confirmed in a study that reviewed the results of various analyzes using the particular index [5]. According to Ozpolat and Akar [11], experimental results show that the match ratio between the obtained learner's learning characteristics using the questionnaires traditionally used for learning style assessment is high for most of the dimensions of learning style.

3 Design of the Study

The participants of the study were students of the Hellenic Open University and were randomly selected. As a result, they vary considerably with respect to their age, background knowledge and skills as well as the reason they need the degree and the time they have to spend. Learning styles may vary considerably among the users of a typical class of students of the same age. Therefore, in a class of HOU the diversity is even greater.

Specifically, 699 users were selected to participate in the empirical study and answer a questionnaire composed by a set of 73 questions. The questionnaire was carefully designed taking into account a wide range of personal characteristics and skills of the students that participated in the empirical study according to the fundamental learning styles identified by the Index of Learning Styles (ILS) [6].

4 Dataset Formation and Standardization

Let

$$S = \{s_1, \ldots, s_M\} \tag{1}$$

be the set of students pertaining to the empirical study described in Sect. 3 and

$$Q = \{q_1, \ldots, q_L\} \tag{2}$$

the aforementioned set of questions such that $M = 699$ and $L = 73$. Taking into consideration the fact that each question $q_j \in Q$ admits a distinct set of possible answers, we may express the domain set for each question as:

$$dom(q_j) = \{a_j^1, \ldots, a_j^r, \ldots, a_j^{n_j}\} \tag{3}$$

given that n_j is the number of different answers for the j-th question. Moreover, by collecting the number of distinct answers per question, the set

$$\mathcal{N} = \{n_1, \ldots, n_L\} \tag{4}$$

can be formed.

In this setting, we could define an injective mapping γ as:

$$\gamma : dom(q_j) \to \mathbb{N}^* \tag{5}$$

such that the following equivalence holds:

$$\gamma(a_j^r) = r \Leftrightarrow a_j^r = \gamma^{-1}(r) \tag{6}$$

where $\gamma^{-1}(\cdot)$ is the inverse function of $\gamma(\cdot)$. That is, each distinct answer a_j^r within the set $dom(q_j)$ can be identified by the corresponding natural number r

and retrieved through the utilization of the inverse mapping γ^{-1}. In view of the previous declarations, we may define the mapping:

$$\mathcal{A} : S \times Q \rightarrow \mathbb{N}^* \tag{7}$$

which actually retrieves the students' answers on each question such that $\gamma^{-1}(\mathcal{A}(s_i, q_j))$ represents the answer of the i-th student to the j-th question. Thus, the formation of the original dataset is conducted by collecting the answer of each student for any given question in a $M \times L$ matrix $\mathbb{A}$ such that:

$$\mathbb{A} = [a_{ij}] = [\mathcal{A}(s_i, q_j)], \forall i \in [M], \forall j \in [L] \tag{8}$$

where $\mathcal{A}(s_i, q_j) \in [n_j]$.

Finally, in order to enforce zero mean and unit variance amongst the column-wise entries of $\mathbb{A}$, the following transformation is being utilized:

$$\hat{a}_{ij} = \frac{a_{ij} - \mu_j}{\sigma_j}, \forall i \in [M], \forall j \in [L] \tag{9}$$

given that:

$$\mu_j = \frac{1}{M} \sum_{i=1}^{M} a_{ij}, \forall j \in [L] \tag{10}$$

and

$$\sigma_j = \sqrt{\frac{1}{M} \sum_{i=1}^{M} (a_{ij} - \mu_j)^2, \forall j \in [L]}. \tag{11}$$

The standardized version of each original matrix element is assumed to be stored in a new matrix $\hat{\mathbb{A}}$. According to the following row-wise expansion of matrix $\hat{\mathbb{A}}$

$$\hat{\mathbb{A}} = \begin{bmatrix} \hat{a}_1 \\ \vdots \\ \hat{a}_i \\ \vdots \\ \hat{a}_M \end{bmatrix}, \tag{12}$$

each $\hat{a}_i \in \mathbb{R}^L$ corresponds to a student educational profile which may be viewed as a point in a L-dimensional space. By projecting these points into a lower-dimensional (i.e. three-dimensional) space through the utilization of a multi-dimensional scaling technique such as Principal Components Analysis we can obtain the original space distribution of the standardized version of the dataset as illustrated in Fig. 1.

5 Artificial Immune Network

The Artificial Immune Network (AIN) is developed as an edge-weighted graph composed of a set of nodes, called memory antibodies and sets of node pairs,

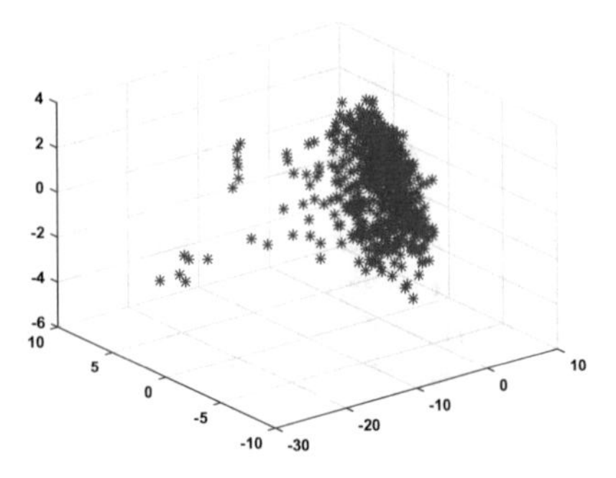

Fig. 1. Original data distribution.

called edges, with an assigned weight or connection strength that reflects the affinity of their match. The antigenic patterns set to be recognized by the AIN will be composed of the set of M L-dimensional feature vectors while the produced memory antibodies can be considered as an alternative compact representation of the original feature vector set.

In order to quantify immune recognition, we consider all immune events as taking place in shape-space $\mathbb{S}$ which constitutes a multi-dimensional metric space where each axis stands for a physico-chemical measure characterizing a molecular shape. For the purposes of the current paper, we utilized a real valued shape-space where each element of the AIN is represented by a real valued vector of L elements. This implies that $\mathbb{S} = \mathbb{R}^L$. The affinity/complementarity level of the interaction between two elements of the AIN is computed on the basis of a correlation-based distance measure between the corresponding vectors in $\mathbb{R}^L$.

Specifically, for a pair of student profiles $\hat{a}_p, \hat{a}_q \in \mathbb{R}^L$, their correlation-based distance will be evaluated through the utilization of the following equation:

$$d(\hat{a}_p, \hat{a}_q) = 1 - |d_{corr}(\hat{a}_p, \hat{a}_q)| \tag{13}$$

where correlation is quantified by the Pearson Correlation Coefficient which is formulated as:

$$d_{corr}(\hat{a}_p, \hat{a}_q) = \frac{\sum_{j=1}^{L}(\hat{a}_p(j) - \hat{a}_{\overline{p}})(\hat{a}_q(j) - \hat{a}_{\overline{q}})}{\sqrt{\sum_{j=1}^{L}(\hat{a}_p(j) - \hat{a}_{\overline{p}})^2}\sqrt{\sum_{j=1}^{L}(\hat{a}_q(j) - \hat{a}_{\overline{q}})^2}} \tag{14}$$

given that:

$$\hat{a}_{\overline{s}} = \frac{1}{L}\sum_{j=1}^{L}\hat{a}_s(j), \forall s \in [M]. \tag{15}$$

Having in mind that the values of the Pearson correlation-related distance measure provided by Eq. 14 satisfy the following inequalities

$$-1 \leq d_{corr}(\hat{a}_p, \hat{a}_q) \leq +1, \tag{16}$$

it is easy to deduce that the affinity measure proposed in this paper according to Eq. 13 varies within the [0, 1] interval. In particular, the affinity measure adopted in this paper results in a symmetric distance metric which assigns equal distance values between pairs of students' educational profiles which have the same degree of positive or negative correlation.

5.1 aiNet Learning Algorithm

The aiNet learning algorithm, in terms of the utilized biological concepts, can be summarized as follows:

1. *Initialization:* Create an initial random population of network antibodies
2. *Antigenic presentation:* For each antigenic pattern do:
   (a) *Clonal selection and expansion:* For each network element, determine its affinity with the currently presented antigen. Select a number of high affinity elements and reproduce (clone) them proportionally to their affinity;
   (b) *Affinity maturation:* Mutate each clone inversely proportionally to each affinity with the currently presented antigenic pattern. Re-select a number of highest affinity clones and place them into a clonal memory set;
   (c) *Metadynamics:* Eliminate all memory clones whose affinity with the currently presented antigenic pattern is less than a pre-defined threshold;
   (d) *Clonal interactions:* Determine the network interactions (affinity) among all the elements of the clonal memory set;
   (e) *Clonal suppression:* Eliminate those memory clones whose affinity with each other is less than a pre-specified threshold;
   (f) *Network construction:* Incorporate the remaining clones of the clonal memory with all network antibodies;
3. *Network interactions:* Determine the similarity between each pair of network antibodies;
4. *Network suppression:* Eliminate all network antibodies whose affinity is less than a pre-specified threshold;
5. *Diversity:* Introduce a number of new randomly generated antibodies into the network;
6. *Cycle:* Repeat Steps 2 to 5 until a pre-specified number of iterations is reached.

6 Experimental Results

The AIS-based clustering mechanism presented in Sect. 5 was applied on the set of standardized student-specific educational profiles described in Sect. 4. Table 1 presents the particular values chosen for the internal parameters controlling the execution of the algorithm throughout the entire experimentation process. Specifically, the value of σ_s which specifies the minimum allowed distance between any given pair of memory antibodies was experimentally selected to be close to the average distance $D_{mean} = 0.88663$ in the given set of antigenic patterns. Moreover, the value of σ_s which determines the maximum allowed distance between any given educational feature vector and the nearest memory antibody was experimentally set to be sufficiently small. Additionally, the value of the *generations* parameter serves as the actual stopping criterion indicating the total number of iterations for the AIN learning algorithm.

Table 1. AIN clustering parameters

AIN learning parameters	
Generations	20
N	5
ζ	10%
σ_d	0.01
σ_s	0.78

The execution of the AIN learning algorithm generates a set of m memory antibodies which is denoted as:

$$\mathbf{Ab}_{\{m\}} = \{\mathbf{Ab}^1_{\{m\}}, \ldots, \mathbf{Ab}^i_{\{m\}}, \ldots, \mathbf{Ab}^m_{\{m\}}\}, \tag{17}$$

by attempting to jointly satisfy the following directives

$$d(\mathbf{Ab}^i_{\{m\}}, \hat{a}_j) \leq \sigma_d, \ \forall i \in [m], \ \forall j \in [M], \tag{18}$$

and

$$d(\mathbf{Ab}^r_{\{m\}}, \mathbf{Ab}^k_{\{m\}}) \geq \sigma_s, \ \forall r, k \in [m] \times [m] : r \neq k, \tag{19}$$

with a minimum amount of violation. The set of memory antibodies produced for this particular parameter selection has a cardinality of $m = 4$ and is graphically represented in Fig. 2. It is easy to deduce that by selecting the value of σ_s to be close to the average correlation distance within the initial dataset, the number of learning style stereotypes produced is equal to the number of fundamental learning patterns identified by the FSLSM.

The set of memory antibodies produced by the AIN learning algorithm impose a partitioning of the students' educational profiles into m clusters (depicted by Fig. 3) such that:

$$\hat{\mathbb{A}} = \bigcup_{i=1}^{m} \hat{\mathbb{A}}_i \tag{20}$$

where

$$\hat{\mathbb{A}}_k \cap \hat{\mathbb{A}}_l = \emptyset, \forall k, l \in [m] \times [m] : k \neq l. \tag{21}$$

Specifically, the individual clusters $\hat{\mathbb{A}}_r$ are implicitly defined through a mapping of the following form

$$C : \hat{\mathbb{A}} \rightarrow \mathbb{N}^* \tag{22}$$

so that $\mathcal{C}(\hat{a}_p) = r$ could be interpreted as assigning the p-th data point to the r-th cluster. As it was previously mentioned, the most intrinsic way to derive the analytical form of the mapping $\mathcal{C}$ is through the utilization of the derived memory antibodies according to the next equation

$$\mathcal{C}(\hat{a}_p) = \arg\min_{i \in [m]} \{d(\mathbf{Ab}^i_{\{m\}}, \hat{a}_p)\} \tag{23}$$

which yields that the r-th cluster can be retrieved by

$$\hat{\mathbb{A}}_r = \{\hat{a}_j \in \hat{\mathbb{A}} : \mathcal{C}(\hat{a}_j) = r\}. \tag{24}$$

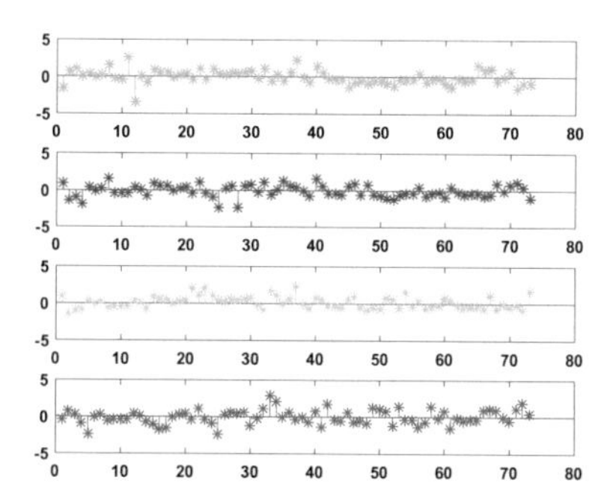

Fig. 2. AIN-Based learning style stereotypes.

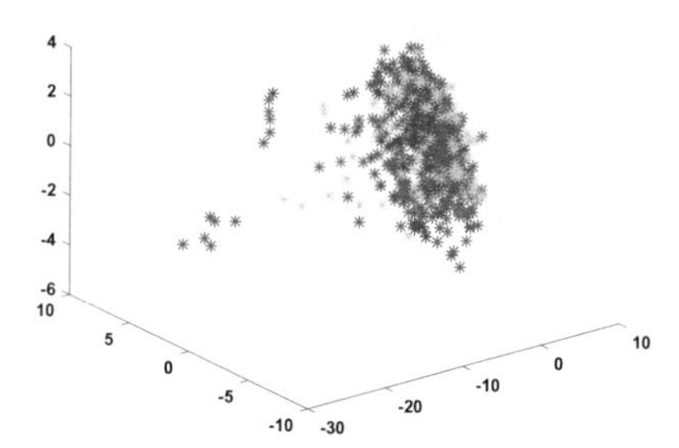

Fig. 3. AIS-Based clustering.

7 Conclusions and Future Work

This paper focused on the problem of extracting fundamental learning patterns from a given set of educational profiles that were collected during the conduction of an extended empirical study. The final product of the clustering technique presented in this paper is a highly diversified and well distributed set of memory antibodies that exhibit a desirable degree of correlation with any given educational feature vector in the dataset. Future work will be focusing on the field of clustering categorical data were traditional algorithmic approaches fail to reveal their intrinsic organization.

References

1. Alfonseca, E., Carro, R.M., Martín, E., Ortigosa, A., Paredes, P.: The impact of learning styles on student grouping for collaborative learning: a case study. User Model. User-Adap. Inter. **16**(3–4), 377–401 (2006)
2. Cassidy*, S.: Learning styles: an overview of theories, models, and measures. Educ. Psychol. **24**(4), 419–444 (2004)
3. Dringus, L.P., Terrell, S.: The framework for directed online learning environments. Internet Higher Educ. **2**(1), 55–67 (1999)
4. Felder, R.M., Silverman, L.K., et al.: Learning and teaching styles in engineering education. Eng. Educ. **78**(7), 674–681 (1988)
5. Felder, R.M., Spurlin, J.: Applications, reliability and validity of the index of learning styles. Int. J. Eng. Educ. **21**(1), 103–112 (2005)
6. Felder, R., Soloman, B.: Index of learning styles questionnaire (1997). Accessed 30 April 2007
7. Garrison, D.R., Kanuka, H.: Blended learning: uncovering its transformative potential in higher education. Internet Higher Educ. **7**(2), 95–105 (2004)
8. Gu, Q., Sumner, T.: Support personalization in distributed e-learning systems through learner modeling. In: Information and Communication Technologies, ICTTA 2006, 2nd edn., vol. 1, pp. 610–615. IEEE (2006)
9. Kolb, A.Y., Kolb, D.A.: Learning styles and learning spaces: enhancing experiential learning in higher education. Acad. Manage. Learn. Educ. **4**(2), 193–212 (2005)
10. Ngai, E.W., Poon, J., Chan, Y.H.: Empirical examination of the adoption of WebCT using TAM. Comput. Educ. **48**(2), 250–267 (2007)
11. Özpolat, E., Akar, G.B.: Automatic detection of learning styles for an e-learning system. Comput. Educ. **53**(2), 355–367 (2009)
12. Rayner, S., Riding, R.: Towards a categorisation of cognitive styles and learning styles. Educ. Psychol. **17**(1–2), 5–27 (1997)
13. Rich, E.: Stereotypes and user modeling. In: User Models in Dialog Systems, pp. 35–51. Springer (1989)
14. Wang, K.H., Wang, T., Wang, W.L., Huang, S.: Learning styles and formative assessment strategy: enhancing student achievement in web-based learning. J. Comput. Assist. Learn. **22**(3), 207–217 (2006)

A Sightseeing Guidebook Automatic Generation Printing System According to the Attribute of Tourist (KadaTabi)

Ryosuke Izumi[1(✉)], Takayuki Kunieda[1], Yusuke Kometani[2], Naka Gotoda[2], and Rihito Yaegashi[2]

[1] Graduate School of Engineering, Kagawa University, Hayashi-cho 2217-20, Takamatsu, Kagawa 761-0396, Japan
s18g456@stu.kagawa-u.ac.jp

[2] Faculty of Engineering and Design, Kagawa University, Hayashi-cho 2217-20, Takamatsu, Kagawa 761-0396, Japan

Abstract. Many sightseeing spots distributes sightseeing guidebooks of paper media at facilities such as tourist information centers, hotels, and stations which are the base of sightseeing. A paper sightseeing guidebook is valuable means to collect sightseeing information for tourists. However, from the tourist's side, it is difficult to quickly find guidebook that one wants because too many kinds of guidebooks are distributed. From the information provider side, if the content of the guidebook changes, it is necessary to collect and correct the once distributed guidebooks and rebuild and redistribute. Also, there is a problem that it is impossible to grasp the use situation of the guide book. We developed the sightseeing guidebook automatic generation printing system according to the attribute of tourists (KadaTabi). KadaTabi is a system that creates recommended sightseeing guidebooks considering tourist attribute and sightseeing season and prints via a cloud printer. In this paper, we describe the development of a sightseeing guidebook automatic generation printing system according to tourist attributes (KadaTabi).

Keywords: Sightseeing guidebook · Sightseeing information · Tourist attribute Local information

1 Introduction

"Tourism Nation Promotion Basic Plan" has been approved by the Cabinet in March 2012. "Tourism Nation Promotion Basic Plan" is the basic plan about tourism nation realization which bases on "Tourism Nation Promotion Basic Low" and tourism measures which bases on "Tourism Nation Promotion Basic Plan" is conducted. "Tourism Nation Promotion Basic Plan" clearly positions tourism as a pillar of Japan's important policy.

Maeda [1] defined sightseeing information as "information which is necessary for tourists in every situation when they sightsee. Yasumura [2] classified tourist information by stage of tourism behavior. Ichikawa [3] describes the classification of tourist information made by Yasumura that sightseeing information is classified into three information; "advance information" which is necessary at the preparatory stage, "field

M. Virvou et al. (Eds.): JCKBSE 2018, SIST 108, pp. 207–213, 2019.
https://doi.org/10.1007/978-3-319-97679-2_21

information" which is necessary at the destination, and "post information" which is handled after sightseeing, and it is necessary to transmit appropriate information with contents and form according to each stage.

One of the "field information" is sightseeing guidebook. Many sightseeing spots distributes sightseeing guidebooks of paper media at facilities such as tourist information centers, hotels, and stations which are the base of sightseeing. In addition, many tourists use sightseeing guidebooks of paper media to obtain information at sightseeing spots. Sightseeing guidebooks are prepared by tourism associations in the area and persons in charge of each facility, and in addition to the fact that the explanation about tourist spots is substantial, photographs and various necessary information for sightseeing are posted. However, from the tourist's side, it is difficult to find the guidebook that he wants quickly because too many kinds of guidebooks are distributed. From the information provider's side, if the content of the guidebook changes, it is necessary to collect and correct the once distributed guidebooks and rebuild and redistribute them. Also, there is a problem that it is impossible to grasp the use situation of the guide book. In many sightseeing spots, recommended sightseeing routes are introduced as model courses. However, model courses are limited in number and combination, and can't take into consideration the transportation method of tourist and staying time.

We developed the sightseeing guidebook automatic generation printing system according to the attribute of tourists (KadaTabi). KadaTabi is a system that creates recommended sightseeing guidebooks considering tourist attribute, sightseeing season and not only sightseeing spot such as pleasure jaunt but also experiential sightseeing spots, and prints via a cloud printer.

2 Development of KadaTabi

This section describes the sightseeing guidebook automatic generation printing system according to the attribute of tourists (KadaTabi). Section 2.1 describes the outline of KadaTabi. Section 2.2 describes generated guidebook printing application. Section 2.3 describes recommendation guidebook generation application. Section 2.4 describes recommendation guidebook generation function. Section 2.5 describes sightseeing spot library. Section 2.6 describes composition of guidebook generated by KadaTabi.

2.1 The Outline of KadaTabi

Figure 1 shows the outline of KadaTabi. KadaTabi application consists of generated guidebook printing application, recommendation guidebook generation application and sightseeing route recommendation server. Generated guidebook printing application has guidebook selection function to select guidebook and guidebook acquisition information registration function to count the number of acquired guidebook into the DB. Recommendation guidebook generation application has user information acquisition function to select tourist attribute and guidebook display function to display the generated guidebook of HTML format. Sightseeing route recommendation server has recommendation guidebook generation function to generate guidebook in consideration of the acquired tourist attribute and visited time, and sightseeing spot library to store data such as bus timetable. Guidebook of PDF format which is generated by each application is printed via a cloud printer by submit function.

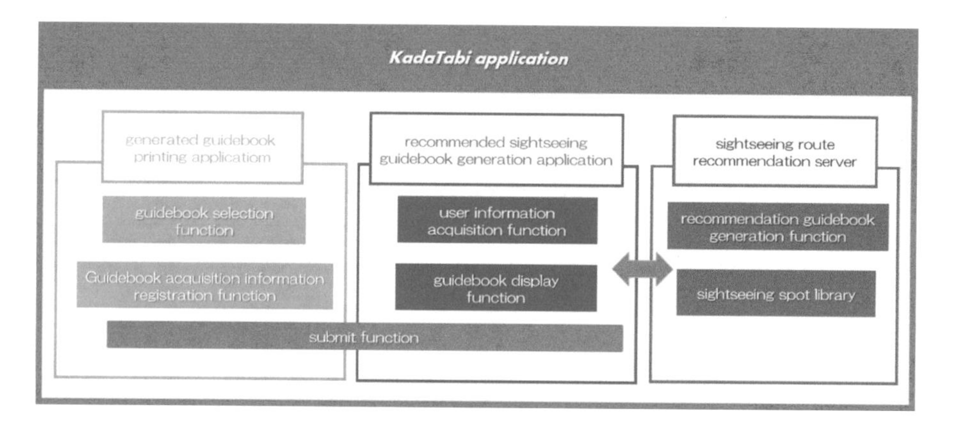

Fig. 1. The outline of KadaTabi

2.2 Generated Guidebook Printing Application

Generated guidebook printing application consists of guidebook selection function and guidebook acquisition information registration function. Guidebook selection function is function that tourists choose guidebook of PDF format which they want from screen of cloud printer or a tablet installed near the cloud printer. If tourists choose images of guidebook which they want, the guidebook is downloaded. When the download is completed and print button is pressed, it is printed. Guidebook acquisition information registration function is function that resisters acquired number and acquired time of each guidebooks into the DB.

2.3 Recommended Sightseeing Guidebook Generation Application

The recommended sightseeing guidebook generation application has a user information acquisition function and guidebook display function. User information acquisition function is a function which tourist chooses own trip and submits the information to sightseeing route recommendation server. Figure 2 shows user information acquisition screen. There are transportation method, duration of stay, sightseeing spots tourist wants to visit and whether or not to wish for experiencing type sightseeing as selection information. In addition, if transportation method is a bus, tourist enters time of going and returning ferries. If tourist wants experiencing type sightseeing, experiencing type sightseeing list is displayed such as Fig. 3. By checking the content which tourist wants to experience, the route which includes the experiencing type sightseeing is recommended. Guidebook display function is a function which displays sightseeing route which is recommended by sightseeing route recommendation server as guidebook of HTML format.

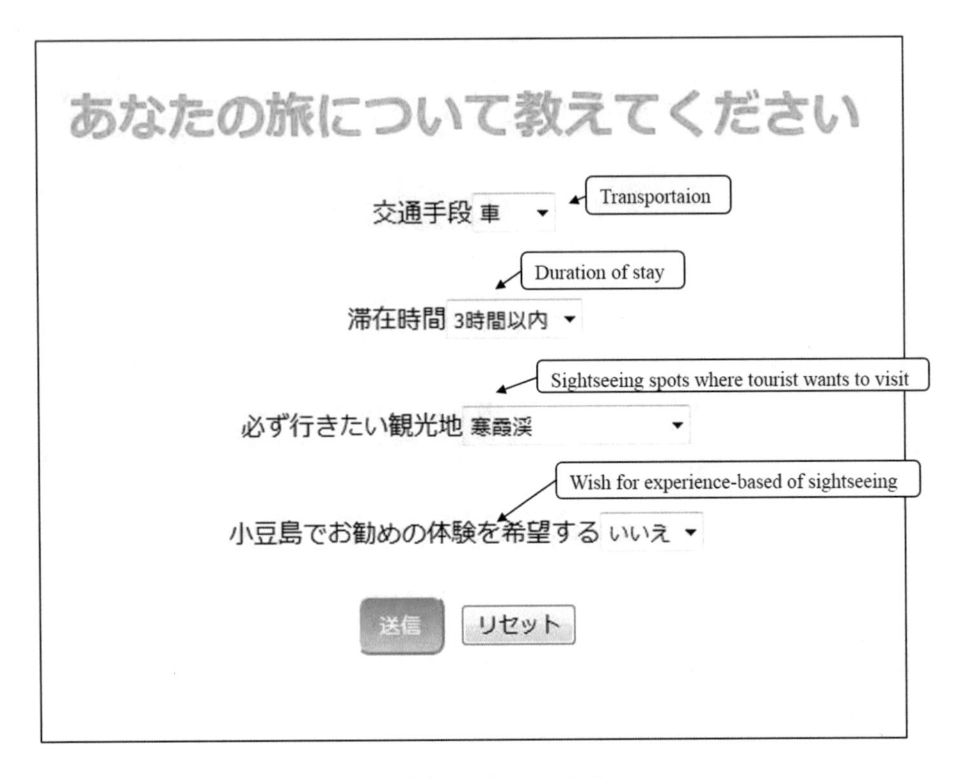

Fig. 2. User information acquisition screen

2.4 Recommendation Guidebook Generation Function

Recommendation guidebook generation function consists of sightseeing spot extraction function, shortest path search function, necessary time measurement function and log data. Sightseeing spot extraction function is a function which extracts three sightseeing spots from sightseeing spot library using information acquired by the user information acquisition function and the random number. If there are sightseeing spots and experiences which tourists want to visit, the number of sightseeing spots extracted with random numbers reduces. Also, sightseeing spots extracted with random numbers may duplicate, so if it duplicates, a random number will be generated again.

Shortest path search function is a function which searches the order which can go around three sightseeing spots extracted with sightseeing spot extraction function in shortest time. There are six ways to go around three tourist spots.It measures the necessary time of six patterns, and the pattern which has shortest time is the shortest path. Necessary time measurement function is a function which searches whether or not total necessary time can fit within the staying time entered by the tourist. Total necessary time is the sum of time to go around the three extracted sightseeing spots and the staying time of each sightseeing spot. If total necessary time fits within the staying time, the route is decided However, if total necessary time doesn't fit within the staying time, another

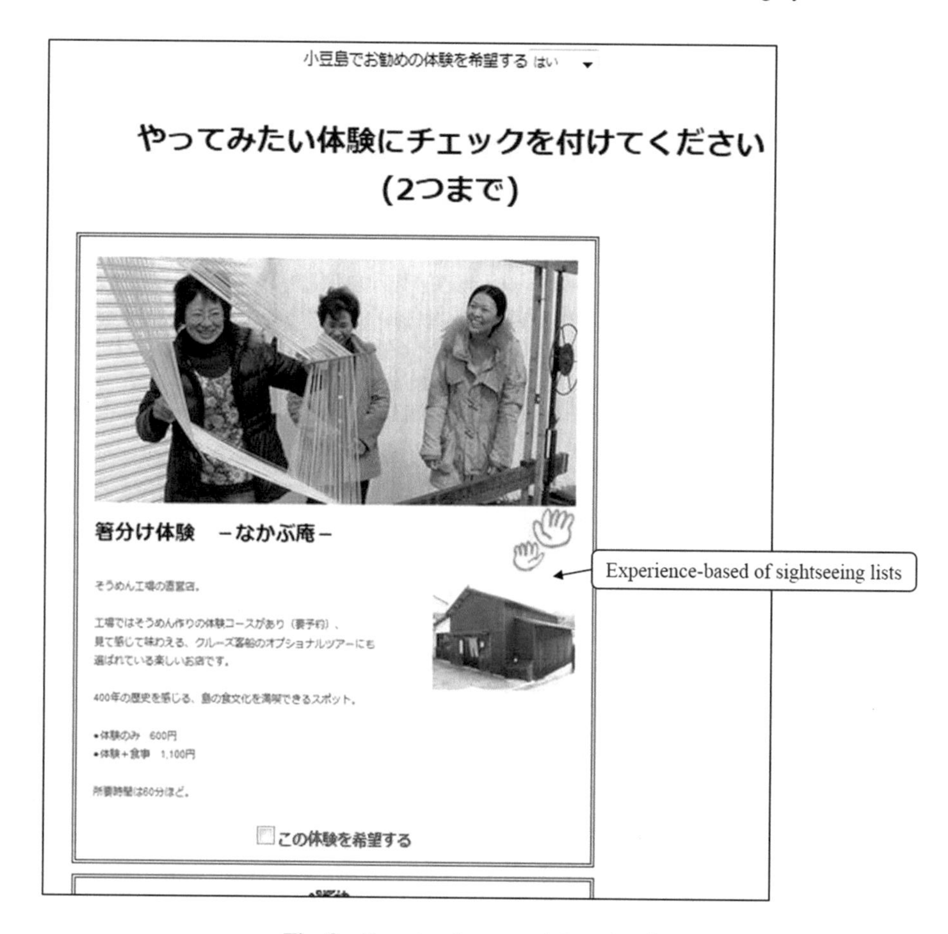

Fig. 3. Experiencing type sightseeing list

sightseeing spot is extracted by sightseeing spot extraction function and repeat the same process until total necessary time fit within the staying time.

The log data acquisition function is a function which registers tourist attribute acquired by the user information acquisition function and information of the tourist spot which route determined into the DB.

2.5 Sightseeing Spot Library

The sightseeing spot library stores sightseeing spot information data, necessary time data between sightseeing spots, the timetable data of bus and log data. The sightseeing spot information data stores basic information of sightseeing spot (sightseeing spot name, longitude, latitude, staying time). These data are used when sightseeing spots are extracted by sightseeing spot extraction function. Necessary time data between sightseeing spots stores the time which takes to travel between tourist spots by car. These data are used when tourists search necessary time between sightseeing spots with shortest path search function. The timetable data of bus stores departure time of bus, bus stop name which are necessary when tourists travel between tourist spots by bus. Log data stores tourist attribute acquired by the user information acquisition function and sightseeing spot information recommended by system.

2.6 Structure of Guidebook Generated in KadaTabi

Figures 4 and 5 show the guidebook generated in KadaTabi. The guidebook is duplex printing and folding in three. Figure 4 shows the surface of the guidebook. A cover is on the right, an explanation of KadaTabi is on the center, a timetable on the left and so on. Figure 5 shows the back side of the guidebook. Information of recommended sightseeing spots, the route connecting between sightseeing spots, and the necessary time are posted on the back side.

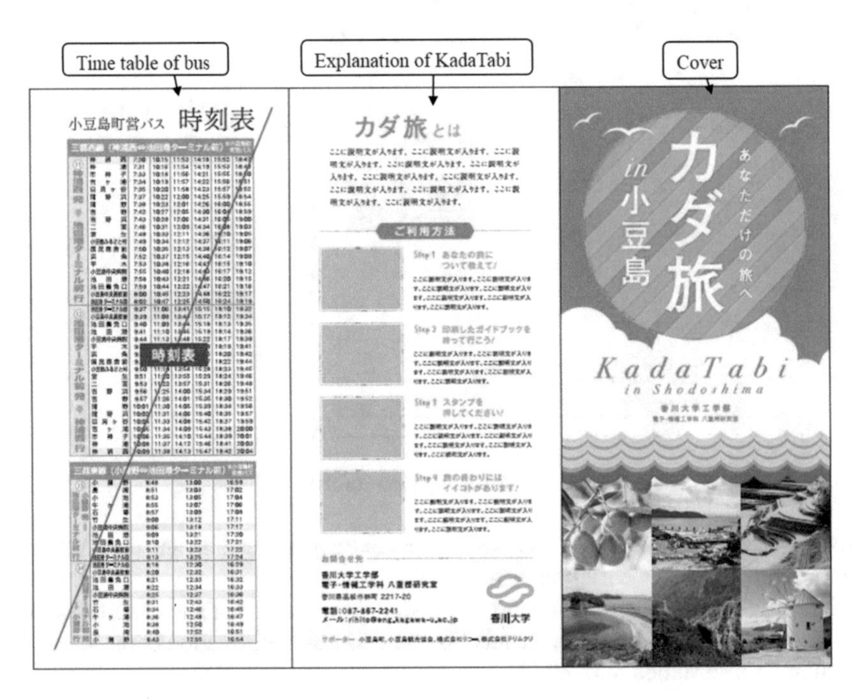

Fig. 4. Guidebook generated by KadaTabi (Surface)

Fig. 5. Guidebook generated by KadaTabi (Back Side)

3 Conclusion

This paper described the sightseeing guidebook automatic generation printing system according to the attribute of tourists (KadaTabi). Tourists can print generated guidebook of PDF format with generated guidebook printing application via cloud printer. The information provider can grasp the use situation of the guidebook and can easily change content of guidebook. Recommended sightseeing guidebook generation application recommends sightseeing spots according to the attribute of tourists and prints via cloud printer. Recommended tourist spots are included not only sightseeing spot such as pleasure jaunt but also experiential sightseeing spots. As future works, the number of recommended sightseeing spots is not fixed at three places but can be changed flexibly according to the attributes of tourists.

Acknowledgment. This research was supported by Ricoh Research Fund, Shodoshima Research Fund, Kagawa University Revitalization Project Expenditure.

References

1. Maeda, I.: Modern Tourism General Remarks. 3nd edn. Gakubunsya (2007)
2. Yasumura, M., Noguchi, Y., Konno, M.: Tourism business argument lectutre on kunpul (2005)
3. Ichikawa, H., Abe, A.: Information technology support on trip around. Jpn. Soc. Artif. Intell. **26**(3), 240–247 (2011)

The Development of the System Which Creates Lecture Contents by a Combination of Various Units and Learning Contents

Tomoki Yabe[1(✉)], Teruhiko Unoki[2], Takayuki Kunieda[1], Yusuke Kometani[3], Naka Gotoda[3], Ken'ichi Fujimoto[3], Toshihiro Hayashi[3], and Rihito Yaegashi[3]

[1] Graduate School of Engineering, Kagawa University, Hayashi-cho 2217-20, Takamatsu, Kagawa 761-0396, Japan
s18g479@stu.kagawa-u.ac.jp

[2] Imagica Robot Holdings Inc., Uchisaiwai-cho Tokyu Bldg. 11F Uchisaiwaicho 1-3-2, Chiyoda-ku, Tokyo, 100-0011, Japan

[3] Faculty of Engineering and Design, Kagawa University, Hayashi-cho 2217-20, Takamatsu, Kagawa 761-0396, Japan

Abstract. We developed the system which creates lecture contents by a combination of various units and learning contents. This system can divide lecture contents into each unit and learning contents by using syllabus metadata generated from the syllabus. This system also can create lecture contents combined with various units and learning contents from systemized lecture information. This paper describes a method to systematize relations between lecture and automatic generation of lecture contents combined with various units and learning contents from systemized lecture information.

Keywords: Lecture contents · Systematization of lecture information eLearning

1 Introduction

We developed the lecture contents viewing system using lecture contents metadata [1, 2]. The system which we developed has three functions, index function, unit/content viewing function and playlist function. The index function can view lecture contents from utterance time of indexing term by using index metadata. The unit/content viewing function can play back selected unit and learning content of lecture content by using syllabus metadata generated from the syllabus. The playlist function can view plural units and learning contents continuously. This system provides various mechanism of viewing lecture contents on learning. However, this system does not have educational functions to indicate suitable units for learners, promote to learn basic units of current one and create lecture contents combined with various units/learning contents. It is because that that system cannot systematize the relation between units and learning contents in lectures.

M. Virvou et al. (Eds.): JCKBSE 2018, SIST 108, pp. 214–220, 2019.
https://doi.org/10.1007/978-3-319-97679-2_22

This paper describes a method to systematize relations between lecture and automatic generation of lecture contents combined with various units and learning contents from systemized lecture information.

2 Lecture Contents Viewing System Using Lecture Contents Metadata

Lecture contents metadata consist of index metadata and syllabus metadata. The index metadata is generated from text data, which is converted by teacher's utterance in lecture contents using a voice recognition technology. Figure 1 shows index metadata. In Index metadata, the information of lecture contents describe contents tags, indexing term describe term tags and utterance time of indexing term describe time tags. Figure 2 shows that the user can directly enter in the input form if the word which user wants to search is clear. Figure 3 shows the result, when it is stored in the database. In order to realize the index system, the contents created by this system don't need the play back from the beginning but from the midstream. The Syllabus metadata is generated from the Syllabus. Figure 4 shows the Syllabus metadata. According to Fig. 4, the lecture contents JAD02 consists of various units and learning contents. This Unit consist of encoding of information and information content. This learning content consist of information transmission and symbol and code. The lecture contents can be played back selected units and learning contents by using syllabus metadata. In this research, the authors use SMIL (Synchronized Multimedia Integration Language) [3] of markup language. Using SMIL, we can create a video combined various multimedia such as moving image data, image data, sound data, and text data. Figure 5 shows SMIL file generated by playback of units and learning contents function. According to Fig. 4, the lecture contents JAD02 playback unit of information content from 1905 s to 3008 s. These information are based on syllabus metadata which the author defined in this research.

```
<contents name="JAD02">
  <index>
    <term name="Encoding">
      <time>00:04:39</time>
    </term>
    <term name="Symbol">
      <time>00:07:32</time>
      <time>00:14:12</time>
      <time>00:46:08</time>
      <time>01:07:29</time>
      <time>01:07:54</time>
    </term>
    <term name="Codeword">
      <time>00:08:03</time>
      <time>00:12:11</time>
      <time>00:46:08</time>
    </term>
    <term name="Length of code">
      <time>00:10:23</time>
    </term>
    <term name="Average code length">
      <time>00:10:37</time>
      <time>00:25:46</time>
    </term>
    …
  </index>
</contents>
```

Fig. 1. The index metadata

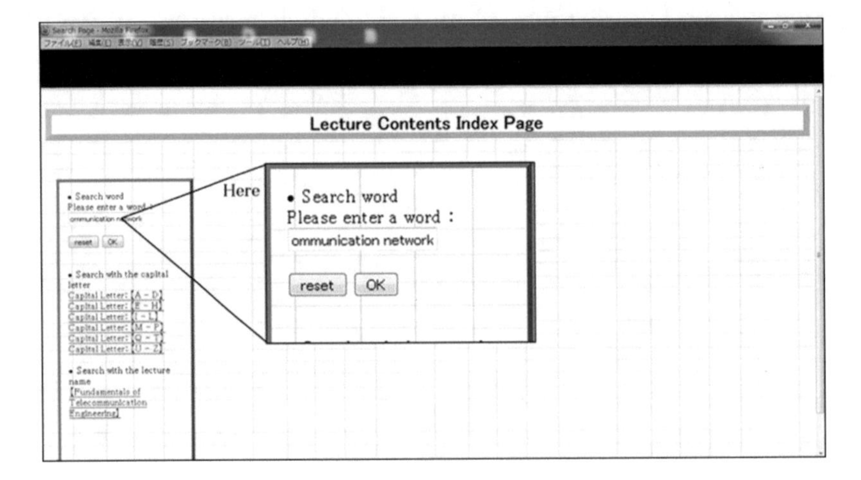

Fig. 2. Search

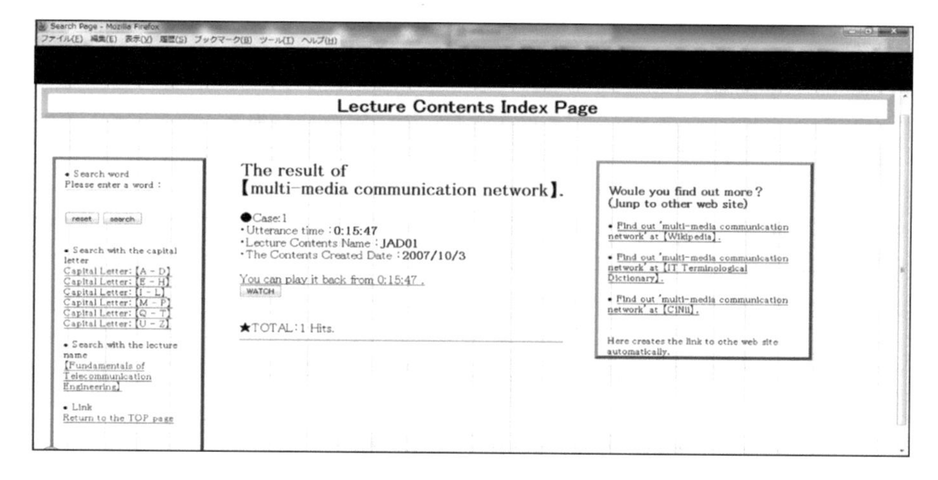

Fig. 3. Result

```
<lecture title=" Fundamentals of Telecommunication Engineering">
  <unit id="1">
    <unitstart>00:00:00</unitstart>
     <unitend>00:31:45</unitend>
     <unittitle>Encoding of Information</unittitle>
     <content id="1">
       <contentstart>00:00:00</contentstart>
        <contentend>00:06:41</contentend>
       <contenttitle>Information Transmission</contenttitle>
    </content>
    <content id="2">
       <contentstart>00:06:41</contentstart>
       <contentend>00:09:57</contentend>
        <contenttitle>Symbol and Code</contenttitle>
    </content>
…
  </unit>
  <unit id="2">
    <unitstart>00:31:45</unitstart>
    <unitend>00:50:08</unitend>
    <unittitle>Information Content</unittitle>
…
  </unit>
…
</lecture>
```

Fig. 4. The syllabus metadata

```
<smil>
    <head>
      <layout>
        <region id="1" width="480" height="320"fit="fill"/>
      </layout>
    </head>
    <body>
      <video clip-begin="1905s"clip-end="3008s"
        src="JAD02.rm" title="Information Content"region="1"/>
    </body>
</smil>
```

Fig. 5. SMIL file generated by playback of units and learning contents function

3 Systematization of Lecture Information

The ontology [4] is defined to be "Explicitly expressing emerged components when viewing the objective world from a certain viewpoint, and summary of systematically those relationships." In this research, relations between lectures were systemized by the ontology. The ontology clearly shows the presence which constructs the world of consideration target, and it can create the base of premise of world understanding by those relationships. In this research, it aims to define the elements constituting the world of lecture in university and systematize those mutual relationships. We use the ontology to systematize those mutual relationships. Figure 6 shows the ontology systematized from information related subjects and contents of these units at Shibaura Institute of Technology by using Hozo – Ontology Editor [5]. "Introduction to Computer Systems" is a lecture, which has been offered at Shibaura Institute of Technology. "Binary number" is the second lecture of "Introduction to Computer Systems". "Fundamentals of Telecommunication Engineering" is a lecture, which has been offered at Shibaura Institute of Technology. "Encoding of the information" is second lecture of "Fundamentals of Telecommunication Engineering". "Binary number" is the necessary prerequisite knowledge to take the lecture of "Encoding of the information". In the similar way, "Encoding of the information" is the necessary prerequisite knowledge to take the lecture of "Information Quantity". Systemized lecture information by using ontology can confirm next learning contents and prerequisite knowledge.

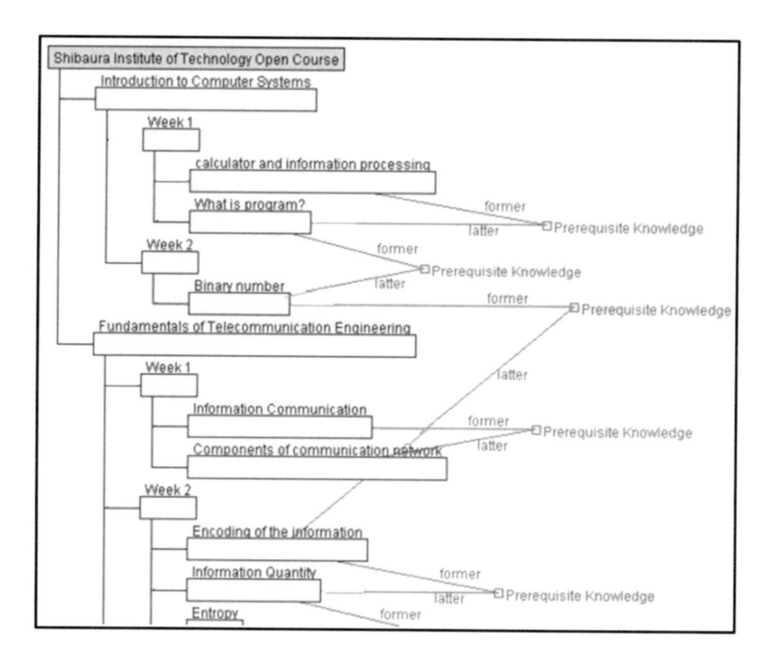

Fig. 6. Subjects held in Shibaura Institute of Technology

4 The System Which Creates Lecture Contents by a Combination of Various Units and Learning Contents

Figure 7 shows the overview of the system which creates lecture contents by a combination of various units and learning contents. The lecture contents are divided into various units and learning contents by the metadata generated from the syllabus. First, the learner selects units and learning contents which they want to learn. Second, the learner creates a new lecture contents from selected units/learning contents by the system, which creates lecture contents by a combination of various units and learning contents. The example of Fig. 7 shows a learner selects unit "a" of lecture content "A", unit "b" of lecture content "B" and unit "c" of lecture content "C". In systematized ontology about relations between lectures, when unit "c" is the precondition of unit "a", and unit "b" is the precondition of unit "a", this system which we developed can create a new lecture content, which can be played back in order of c, a, and b. In this system, by using the lecture information systemized by ontology, lecture contents can be generated considering the order to be learned.

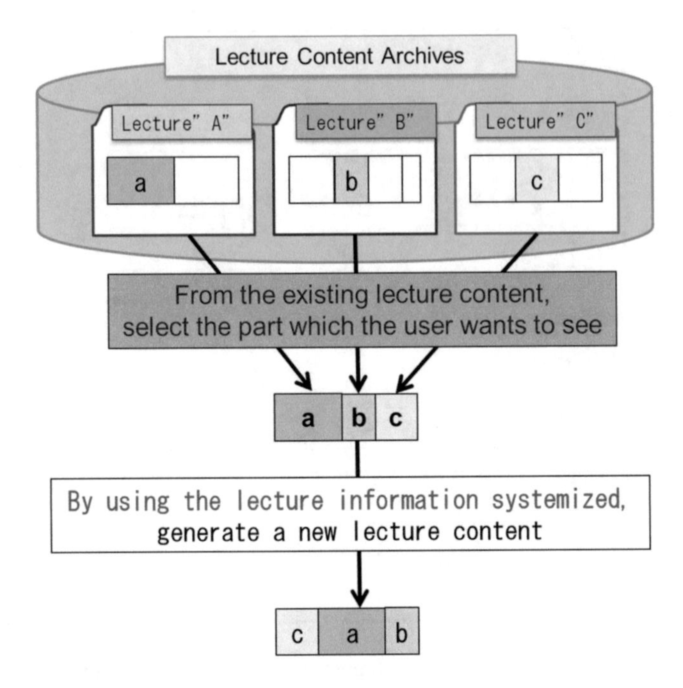

Fig. 7. Overview of the system

5 Conclusion

This paper described automatic generation function by using systemized lecture information with ontology. The learners can confirm the lecture contents which the learner should learn to do next and prerequisite knowledge. In addition, this system can create the new lecture contents to match the learners by using lecture contents combined with various units and learning contents.

Acknowledgment. In promoting this research, I would like to thank everyone at Imagica Robot Holdings Inc., and Photron Limited., for providing equipment and technical support.

References

1. Saitoh, T., Hayashi, T., Yaegashi, R.: The lecture contents with index for self study and its system. In: Proceedings of International Conference on Information Technology Based Higher Education and Training 2012 (ITHET 2012), 6 pages (2012)
2. Saitoh, T., Hayashi, T., Yaegashi, R.: Self-study support system using the lecture contents: creation of study ontology from syllabuses. In: Proceedings of International Conference on Information Technology Based Higher Education and Training 2012 (ITHET 2012), 6 pages (2012)
3. Synchronized Multimedia Integration Language (SMIL 3.0). http://www.w3.org/TR/SMIL/
4. Mizoguchi, R.: Science of intelligence ontological engineering, Ohmsha (2011)
5. Hozo. http://www.hozo.jp/. Accessed 17 May 2018

Development Case of Information Services to Accelerate Open Innovation and Implementation

Through Demonstration Experiments in Kagawa

Takayuki Kunieda[1(✉)], Yusuke Kometani[2], Naka Gotoda[2], and Rihito Yaegashi[2]

[1] Graduate School of Engineering, Kagawa University, Hayashicho 2217-20, Takamatsu City, Kagawa Prefecture 761-0396, Japan
s17d451@stu.kagawa-u.ac.jp

[2] Faculty of Engineering and Design, Kagawa University, Hayashicho 2217-20, Takamatsu City, Kagawa Prefecture 761-0396, Japan
{kometani,gotoda,rihito}@eng.kagawa-u.ac.jp

Abstract. Diversification of customer's demand, speeding up development, cost saving, globalization and the world are changing. Under such circumstances, open innovation is required in the creation and development of information services. Kagawa University has developed several information services cooperating with universities, companies, and administrations so far. In this paper, we introduce three information services developed against administrative issues in collaboration with Kagawa University and Ricoh Co., Ltd. Then, we explain the necessity of open innovation, development method necessary to realize open innovation, importance of demonstration experiment.

Keywords: Open innovation · SDK · Cloud service
Demonstration experiment

1 Introduction

The creation of innovation is a challenge for many companies. It has become difficult with "closed innovation" which relies solely on our own resources that had been done in many companies until now. Due to customer diversification, product development speed up, globalization. "Open innovation" that collaborates with universities and other companies and creates innovation by widely adopting external technologies and ideas has become important.

From the viewpoint of diversifying customer needs, the situation that the market will be accepted if a large amount of products with the same function as in the past economic growth period were developed at low cost mainly decreased in developed countries, the same situation Even for products, the demands of individual customers have diversified and customization for each customer and response to diversified needs has not been accepted in the conventional way.

M. Virvou et al. (Eds.): JCKBSE 2018, SIST 108, pp. 221–230, 2019.
https://doi.org/10.1007/978-3-319-97679-2_23

The speed of product introduction to the market has been accelerated more and more and it is impossible to follow the product development speed of venture companies and other companies by grasping the needs of customers and grasping the needs of customers that large enterprises have taken, It came to be a situation where the product was not used already at the stage of introducing it to the market.

As in the past, product development for a single domestic market alone failed to expand the market, and in the circumstance that many products flow from overseas, global response has become an important viewpoint for companies survival.

Under such circumstances, with the product development as it is, it is getting into a state where it cannot be compared. The creation of innovation through "open innovation" has been drawing attention as a new way to find a way of life by changing the way we have done so far.

The "open innovation" activities that cooperate with industry, government and academia, which will speed up the creation of new customer values will become increasingly important from now on. Kagawa University created information services leading to innovation in collaboration with municipalities and enterprises, and has conducted development and market verification. We introduce the approach to realize open innovation as an example of the information service developed so far.

We also introduce the development method we have done to realize these information services and describe the market verification and its result to judge the effectiveness of the developed information service.

In this paper, Sect. 2 will describe the necessity of open innovation and its background. Section 3 introduces three information service cases developed by Kagawa University. Section 4 describes the creation, development and market verification of these information services. Conclusion discusses the effectiveness of open innovation and the current issues.

2 Necessity of Open Innovation and Its Background

In order to create a new information service by open innovation approach and to develop a system, conventional developing methods cannot follow up, and a new development method is needed. We examined what kind of approach is necessary from the viewpoint of software development.

In software development represented by waterfall type development, in order to develop high-quality and high-reliability software, a method of confirming the request in the early stage of development and advancing development by specifying the request. Due to the rapid development of information and communication technologies, customer needs have diversified, and in order to develop competitive software, it is different from waterfall type development so far, such as agile development mainly focusing on web applications and web system has been proposed as a development method. Agile development refers to a development method that attempts to minimize risk by adopting development in a short time period called iteration. In Agile development, in order to develop software in a short period of time, it is common to develop it using Software Development Kit (SDK) which is a package that required programs and documents are packed together.

In recent years, due to speeding up of networks including mobile communication and performance improvement of mobile terminals, SDKs that can be used via the network have appeared and widely used. Along with that, various information services utilizing the SDK are created. Increase not only business methods that provide products and services based on their own technology called closed innovation but also companies using a technique called open innovation that integrates external knowledge and technology into their own technologies to develop new products are doing.

Ricoh Co., Ltd., a joint research with Kagawa University, will provide many developers with the functions of their own technology and products so as to accelerate open innovation so that even if there is no technical knowledge. We are making efforts to create various information services using technology and to use our own technology through information service created by many users.

Many companies in Japan pursue closed innovation, in which all monopolies and profits of technologies with high competitive superiority are still returned to their company, but there are demerits that require huge cost and time from research and development to commercialization is there [1]. As global competition intensifies, the number of companies aiming to break away from this closed innovation is increasing. Figure 1 represents a model in which information services are created using various SDKs.

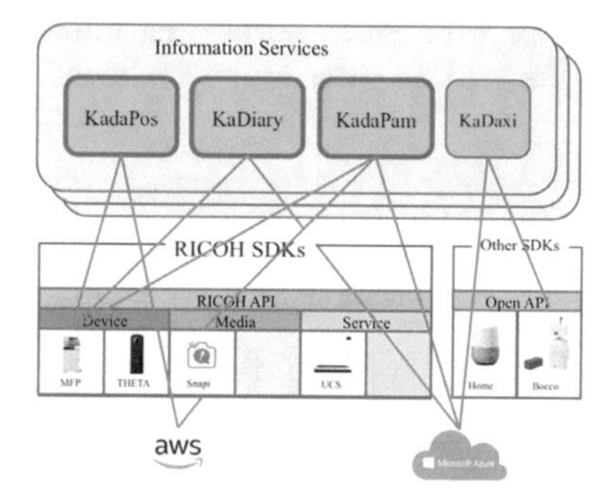

Fig. 1. Provision of SDK and its effect

3 Introduction of Information Service Cases Developed by Open Innovation

In this section, we will introduce examples of three information services that Kagawa University, Ricoh Company, Kagawa Prefecture and Shodoshima Town developed in collaboration and actually carried out demonstration experiments.

3.1 Development of a Advertisement Printer System (KadaPos) [2]

Introduction of KadaPos

The advertisement printer system "KadaPos" is a system that provides regional information to students by printing the area information according to the attributes of the students on the backside of the printing paper. Students can use "KadaPos" for free.

The regional information means information on local shopping districts, information on events held in the area, information on local professional sports organizations and etc. KadaPos consists of a multifunction printer and a KadaPos server built in the cloud. The KadaPos server consists of a print server, an advertisement insertion server, and a user authentication server. In order to advance the development of this information service (KadaPos), Ricoh Co., Ltd. has been improving each stage of the print control mechanism.

Phase 1: Direct Printer Operation

In Phase 1 of KadaPos, we developed a system by directly controlling the Multi-function Printer provided by Ricoh. As a result, it took considerable time for development and cost to repair after release. The reasons are as follows:

1. API for MFP operation provided by Ricoh for this development is not normally open to the public, but it is for internal use.
2. Changes in API specifications and functional specifications that are provided are frequently performed according to internal changes and repair, so it is difficult to master how to use, and it takes time to follow the modification.
3. There is no coordination of certification functions, and it is necessary to prepare certification function unique to KadaPos.

Therefore, in order to use the MFP when developing information services, it is necessary to develop it under the guidance of a designer who is familiar with the control of the MFP (Fig. 2).

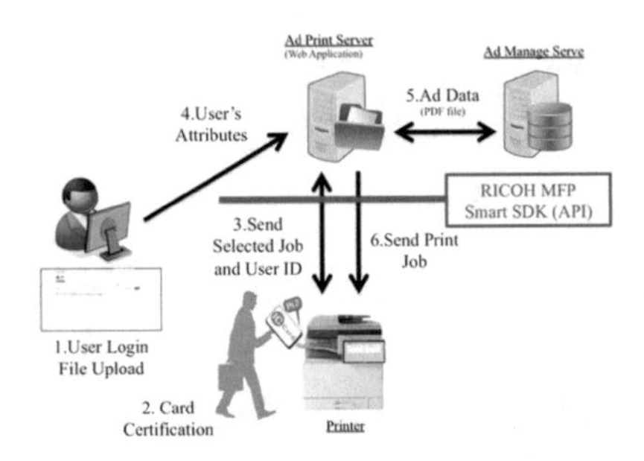

Fig. 2. System configuration of KadaPos 1

Phase 2: Cloud Print

In the conventional MFP control, the MFP is directly operated via the network. Therefore, it was necessary to install a server for controlling the MFP when constructing an information service for operating the MFP. In order to construct such an environment, it is necessary to introduce the cost of preparing the server and software for controlling the MFP on the server. Without knowledge and development experience on MFP control, it was not easy to build a system. All functions for operating the MFP are consolidated on the cloud. Information system developers can easily perform print operations by simply sending jobs to the MFP connected to the Internet and the server on the cloud. In cloud print, print processing is performed by using RICOH SmartSDK JavaScript API provided by Ricoh Company, Ltd. KadaPos Phase 2 realized KadaPos using this cloud print mechanism provided by Ricoh. Figure 3 is a block diagram of KadaPos constructed using cloud print.

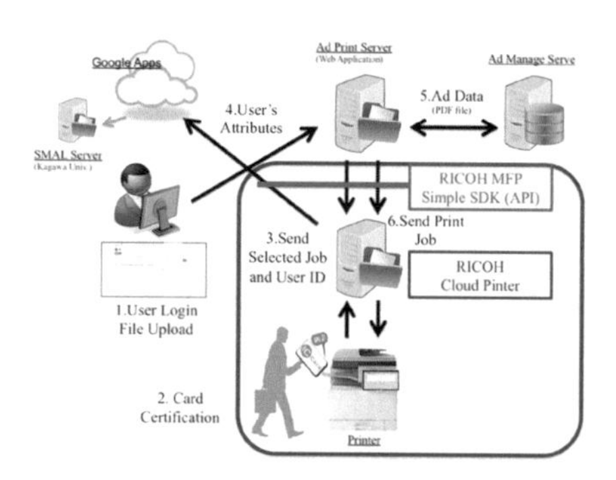

Fig. 3. KadaPos 2 using cloud print

The development efficiency is improved with cloud print, just by developing the system using the API to control the MFP on the cloud. If you understand the API specifications it will be easier to develop the system. MFP can also be used anywhere, anytime, simply by connecting to the Internet. As in the past, it is no longer necessary to install MFP on a specific network and prepare a server to control it. However, it was necessary to understand the detail specifications of the API on the cloud properly, and it did not reach the level of "Easy for everyone".

Phase 3: Web-Based Printing

With the provision of cloud print, information service development has become possible to develop printing system simply by understanding cloud print API specification. The level of the developer is set to a level at which HTML can be described and a website and the like can be developed. By using this function called "Device Tags", users can control the MFP as same as creation of homepage. Device Tags is a Web platform that can create scan/print applications of MFP only by describing the standard (HTML 5). Although the functions to be provided are limited, the usual printing operation can be realized by using this Device Tags. The evaluation is now finished and it is preparing for implementation. The print request is described as follows.

```
<body>
  <a href="http://xx.xx.xx.xx/files/Sample.pdf" download="Sample.pdf">PRINT</a>
</body>
```

KadaPos began demonstration experiments at Kagawa University from January 2016 until the end of July. Effectiveness of KadaPos was seen from usage information acquired during the period. KadaPos continues to be used as a print service within the university after verification. Currently, we are carrying out renovation for Phase 3, and we plan to resume operation after renovation (Fig. 4).

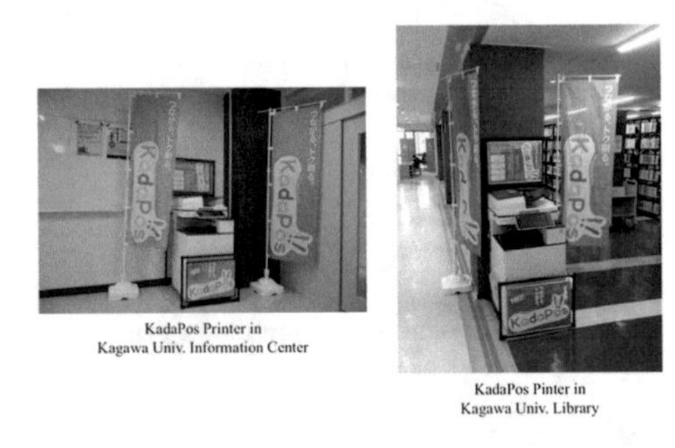

Fig. 4. Installation situation of KadaPos of Kagawa University

3.2 Development of Travel Diary Generating/Printing System (KaDairy) [3]

KaDairy prints and provides the result of tourists traveling around. We can not specify where to place the printers on the tourist spots. In demonstration experiments on Shodoshima Island, we set up a printer at "Michi-no-Eki" (Road Station). Due to the nature of KaDairy, the requirement for the print function is that the installation does not specify the location, it is easy to install, it is necessary to print anywhere at any time. For that reason, we adopted the cloud print adopted for Phase 2 of KadaPos for this print function (Figs. 5 and 6).

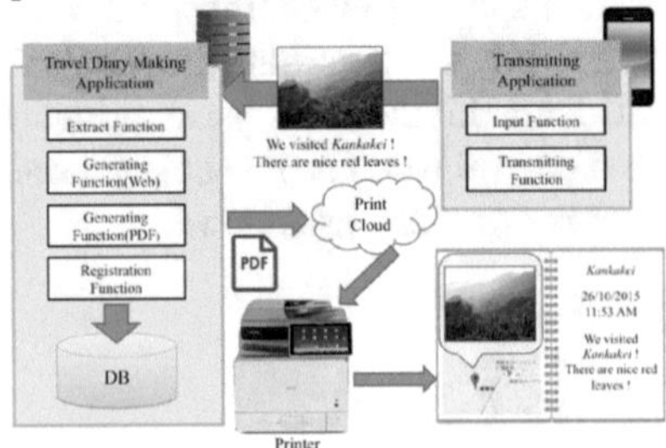

Fig. 5. System outline of KaDariy

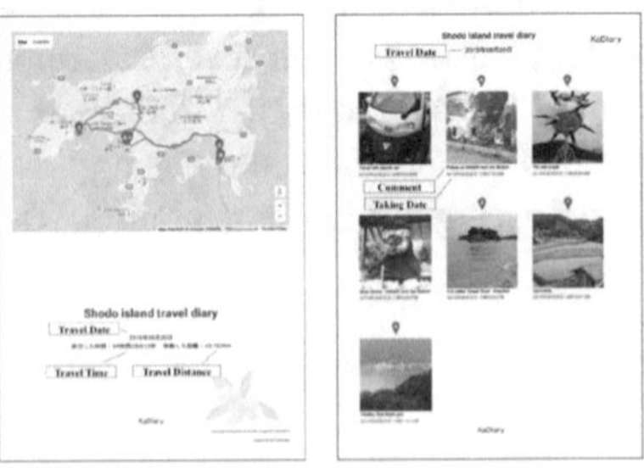

Fig. 6. Sample of Printed Travel Diary

3.3 Development and Tourist Guidebook Generating/Printing System (KadaPam) that Records Memories and Analyzing Travel Behavior

We developed a tourist guidebook generating/printing system (KadaPam [4]) that records travel memories,and we conducted a field experiment in Shodo-shima, Kagawa Prefecture. KadaPam generates a tourist guidebook by replacing the original guidebook images with photos taken by a tourist during tourism.

In this information service, Ricoh Co., Ltd. Print Function (Device Tags) was used. An image recognition technology (RICOH TAMAGO Snapi SDK) [5] was used to determine whether the same object is included in the image taken by the tourist for the set of registered images. This information service was developed as a web-based information service.

It is important that the proposer of the information service conceived the concept of KadaPam. The proposer does not have knowledge of print control method and image recognition technology. Therefore, we must be able to develop the entire system with technology that can develop web-based applications. In order to develop KadaPam, Ricoh Co., Ltd. offered web-based print (Device Tags) to be adopted for Phase 3 of KadaPos. Snapi SDK, an image recognition technology, can be easily incorporated into the system by providing it with JavaScript. As a result, KadaPam was completed in a period of about one month (one man-month) by a single person as a web application which the student of the graduate school planned, designed, developed. The time taken to incorporate the Snapi SDK during the development period was about one week (Figs. 7 and 8).

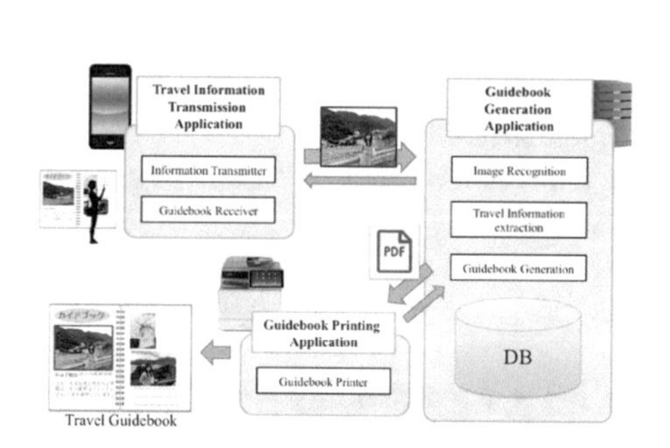

Fig. 7. System overview of KadaPam

Fig. 8. Original guidebook and KadaPam guidebook

We introduced three mechanisms for users who invented information services to easily use technology provided by enterprises. The time chart of Fig. 9 shows how each case evolved.

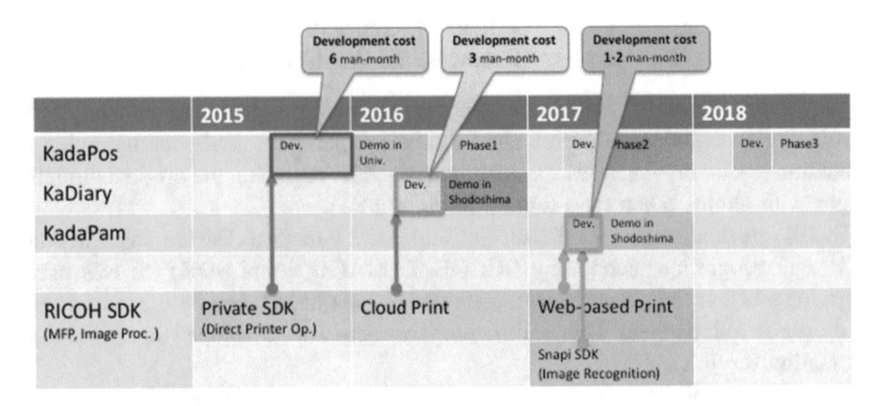

Fig. 9. Time chart of three case studies and transition of development cost

4 Creating, Developing and Marketing Information Services Using Open Innovation

We summarize the ideas necessary to accelerate open innovation from the three information services introduced in Sect. 3.

4.1 Understanding Issues and Creating Ideas

The creation of ideas for each information service is based on the issues of Kagawa prefecture, Takamatsu city, the theme of sightseeing in Shodoshima-cho, etc. We examined what kinds of services are required as information services to solve those problems.

In KadaPos, we propose an information service to solve the problem "I want to revive the decline of the shopping mall in Takamatsu City."

KaDairy proposed an information service to solve the problem "I can not grasp the behavior of tourists visiting Shodoshima Island."

KadaPam proposed an information service to solve the problem that "tour guide books are often thrown out after travel but provided what they left as their own memories".

In this way, it is important for researchers and students to firmly grasp the problems of administrative and tourist spots, to consider solutions with "Ideathon" and to propose appropriate information services.

4.2 Improve Development Efficiency by Exposing the SDK and Simplifying the API

In order to realize open innovation, activities are underway to develop new products by fusing external knowledge and technology to their own technologies. Ricoh Co., Ltd. was established in 2010 TAMAGO Labs. [6, 7]. By the way, TAMAGO is the meaning of eggs in Japanese.

TAMAGO Labs. Is proactively disclosing new technologies to implement open innovation based on the concept of "TAMAGO (Egg) of new business", not only in its own company but also in other industries such as other companies, universities, local governments, technologies possessed by different fields, Ideas, etc., and to develop products based on the needs of customers.

Ricoh Co., Ltd. has prepared the SmartSDK JavaScript API to operate the MFP. From TAMAGO Labs., we provided image recognition technology in the form of Snapi SDK. By providing image recognition technology of Ricoh Co., Ltd. to many developers, Snapi SDK creates various information services using that technology even if there is no knowledge about image recognition technology.

It is important to accelerate open innovation by deciding API specifications as SDK of technology of universities and companies. It is also important to adopt a general-purpose description language such as HTML and make it easy for anyone to use the function.

4.3 Demonstration Experiment Validation and Improvement

Demonstration experiments are important for verifying the effectiveness of the functions provided by the information service and the technology to be used in the developed information service. At Kagawa University, all of the created information services (KadaPos, KaDairy, KadaPam) were actually subjected to demonstration experiments and their effectiveness was evaluated. Through the demonstration experiments, improvement points of the information service itself became clarified, and at the same time it was possible to evaluate not only the function but also ease of use and ease of incorporation of the technology issues. By doing this, we can make improvements to more useful functions and accelerate open innovation.

5 Conclusion

Through the development and demonstration experiments of three information services (KadaPos, KaDairy, KadaPam) so far, the advantages of open innovation became clearer. The creation of information service, improvement of development efficiency and provision service effectiveness for the task can be verified surely by conducting a demonstration experiment.

Especially for development efficiency improvement, it became clear that by providing SDK, which everyone can use easily, it could flexibly deal with the following points:

- Development speed
- Renovation work
- Version upgrade support.

By conducting the demonstration experiments, we were able to confirm the following operational issues as soon as possible for this operation.

- Portability
- Scalability
- Continuity.

Furthermore, with the provision of SDK, we confirmed the following points from the user (developer) point of view by carrying out the demonstration experiment.

- Easy to Use
- Manual less.

In the future we will accelerate the creation of information services by providing various functions as SDK. In addition, users (developers) who develops mash up APIs of various functions and actively use them, so that they will construct information services created by themselves in a short period of time with little effort.

Acknowledgment. In promoting this research, I would like to thank everyone at Ricoh Co., Ltd. for providing equipment and technical support, and everyone on Shodoshima for demonstration experiments.

References

1. Open Innovation. https://bizhint.jp/keyword/12540/. Accessed 10 Apr 2018
2. Takada, R., et al.: Development of a advertisement printer system (KadaPos). Digit. Pract. **8** (4) (2017)
3. Kumano, K., et al.: Development of travel diary generating/printing system (KaDiary). Digit. Pract. **8**(4) (2017)
4. KadaPam. http://www.kadapam.xyz/. Accessed 3 Nov 2017
5. RICOH TAMAGO Snapi. https://www.ricoh.co.jp/software/tamago/snapi/. Accessed 10 Apr 2018
6. TAMAGO Labs. https://www.ricoh.com/software/tamago/. Accessed 10 Apr 2018
7. Kunieda, T.: Challenge to "making new products" in companies. In: Proceedings of the 2017 IEICE Society Conference, B1-3-3. IEICE, Tokyo (2017)

Automated Decision Making and Personal Data Protection in Intelligent Tutoring Systems: Design Guidelines

Eirini Mougiakou, Spyros Papadimitriou(✉), and Maria Virvou

University of Piraeus, Karaoli & Dimitriou St. 80, 18534 Pireaus, Greece
{imougiakou,spap,mvirvou}@unipi.gr
http://www.cs.unipi.gr

Abstract. Paper aims to combine Informatics with Legal Science, in mapping specific areas of conflict between personalized educational applications and GDPR, addressing shortfalls and providing applicable solutions, in the form of guidelines to developers. Attention is focused on Intelligent Tutoring Systems (ITS) under the light of specific GDPR requirements for lawful and transparent Automated Profiling and Automated Decision Making. For this purpose, a self-designed Fuzzy Logic equipped ITS paradigm is utilized as test bed, facilitating the assessment, testing and passing of theory into practice.

Keywords: GDPR · Personal data protection
Personal data processing · Automated decision making
Automated profiling · Consent · Privacy · Right to be informed
Fuzzy logic · Artificial intelligence · Intelligent tutoring systems · ITS
Guidelines · Personalization · Adaptivity · Educational systems
Membership functions

1 Introduction

From everyday surfing via search engines, to networking via social media, users enjoy the benefits of customized experience due to underlying personalization rules. However, few understand the underlying process and fewer the risks entailed. Customization rests on modelling. Modelling rests on personal data analysis using artificial intelligence techniques. Personal data collected has and will continue being object of mistreatment or unauthorized exploitation. This also applies on Intelligent Tutoring Systems (ITS).

ITS applications may utilize artificial intelligence and machine learning in order to address user in a personalized way [1], by appropriate tuning according to student's specifics [2]. In addition, by including adaptive elements [3] in terms of educational context, they improve student's acceptance as learning means, motivation over their usage [4] and ultimately engagement. As such ITS are rendered optimal for personal data protection research, especially under the light

M. Virvou et al. (Eds.): JCKBSE 2018, SIST 108, pp. 231–241, 2019.
https://doi.org/10.1007/978-3-319-97679-2_24

of the European Commission's General Data Protection Regulation (GDPR) [5]; GDPR, propelled by the increase of on-line personal data sharing, aims to enhance data subjects control. Rendered feasible by regulating the collection, processing and storage of personal data, GDPR applies from 25.5.2018 throughout EU, affecting EU residents (both EU citizens and EU visitors) for goods or services provided by both EU as well as non-EU established entities.

The following analysis combines Informatics with Legal Science under a twofold mandate; firstly, mapping shortfalls between ITS applications and GDPR, in terms of automated decision making and profiling; secondly providing applicable solutions, in the form of guidelines to developers [6]. A self-designed and implemented ITS, enhanced with adaptive features by utilizing data mining and analysis, facilitates authors effort in passing theory into practice.

2 Related Work

2.1 Personal Data Protection

Past related publications mostly concentrated on analyzing the legal framework prior to GDRP. Areas assessed included fair principles and best practices employed by large organizations and enterprises, deriving from GDPR-preceding Data Protection Directive 95/46/EC (DPD) [7].

Current publications have shifted towards GDPR, although by the time of writing, a limited number of GDPR related publications addressing automated profiling and decision making is noted. According to the European Union Agency for Network and Information Security (ENISA) [8], privacy consideration can be effectively addressed primarily by user's own knowledge over online privacy [9]. Neither purpose build applications, nor providers best practice can substitute self-awareness.

Facilitating data subjects' effort to achieve self-control over their own personal data, Wachter, Mittelstadt and Floridi [10] argued on the need of a 'right to explanation', applying on all automated decision-making, thus enhancing both accountability and transparency. Unfortunately, by reviewing GDPR early drafts, the right to explanation seems to have been intentionally removed in favour of the right not to be subject to automated decision-making (per binding Article 22 requirements and non-binding Recital 71 provisions) and the right to access (per binding Article 15 requirements and non-binding Recital 63 provisions) which initially read as a supporting factor, but actually serve as a complicating element - especially when cross-assessing the legally binding requirements of Article 22 with the non-binding provisions of Recital 71.

Goodman and Flaxman [11], document how Article 22, relating to automated decision-making and profiling could set at risk the infrastructures of companies in credit extension business, insurance, computational advertising, social networks by potentially challenging the underlying algorithm, AI and machine learning set up, in favour of new interpretable and comprehensible standards.

2.2 Personalization/Adaptivity in Educational Systems

Educational systems apply on students of different characteristics and needs [12]. This heterogeneity drives the need for personalization and adaptivity, as stressed by researchers [13]. Under this context, identifying the Student's knowledge as well as projecting areas for improvement, both regarding the pre-defined subject, stands as a pre-requisite for tuning the content appropriately.

Problems that contain non-measurable aspects, such as level-of-knowledge, can be addressed by Artificial Intelligence (AI). Answers originate from a decision tree or similar structure. Rules, information triggering aforementioned rules and the outcome may be included in the system's reasoning. Rule-based AI technology is the basis of Fuzzy Logic [14]. Fuzzy rule-based system can effectively support complex systems modelling, by using (a) linguistic variables which represent fuzzy sets and logical connectives of these sets and (b) basic properties and operations defined for fuzzy sets.

Subsequently, a sufficient number of adaptive educational applications successfully utilize fuzzy logic during level-of-knowledge assessment [15]. As such, paper in hand examines GDPR in the context of a student-modelling-equipped ITS that authors have developed and considered as an ideal platform for addressing areas of conflict between personalized education and GDPR provisions.

3 Architecture of ITS

The system used as a test bed for the purpose of the paper in hand, consists of an ITS web application openly accessible to internet users. System provides HTML training, through a schedule divided between distinct sessions, like HTML page layout, text formatting, tables and forms. Upon enrolment, (a) Student may practice within distinct sessions or through-out the full HTML training schedule, (b) System can maintain and process Student's data and subsequently personalize training according to individual training needs.

Needs are defined according to Student's knowledge level. Knowledge level evaluation is based on Fuzzy Logic. Specifically, every time a Student participates in an application's test, Students performance (success rate) feeds a Fuzzy Reasoner, which transforms information input to knowledge level evaluation output. Output is stored on a database and used in tuning questions' difficulty level, to challenge the Student during the next test session. As such, System provides targeted assistance, addresses knowledge gaps and eventually propels subject's comprehension to the next level.

Student's knowledge level cannot be defined quantitatively and evaluated distinctively. Though Student may have full or no command of the session's context, sometimes this is neither precise, nor clear. Problems including uncertainty and inaccurate data, can be successfully addressed by Fuzzy Logic. Fuzzy Logic theory is employed by Systems Fuzzy Reasoner, which utilizes four fuzzy sets to classify each Student according to Student's knowledge level [15], specifically between Unknown, Unsatisfactory Known, Known, Learned set. Each fuzzy set is described by a membership function (for depiction purposes Un, UK, K and

L respectively), which receives a success rate percentage of a test and returns a value ranging between 0 and 1. The output of all membership functions always sums to 1, that is $\mu_{Un} + \mu_{UK} + \mu_K + \mu_L = 1$.

In case of a membership function return valued as 1, then Student is classified entirely (100%) within the correlating fuzzy set, in other words Student's specific knowledge level is rendered certain. In case of function return being positive but less than 1 (for example 0.6), then Student falls by 60% within a specific set and by 40% within another adjacent set. Student cannot fall simultaneously into Unknown and Learned sets. Specific fuzzy logic characteristic simulates human intelligence, capturing the element of uncertainty, usually present during human evaluation of concepts such as Student knowledge level. System employs the membership functions depicted in Fig. 1.

"Unknown"

$$\mu_{Un}(x) = \begin{cases} 1, & 0 \le x \le 45 \\ 1 - \frac{x-45}{50-45}, & 45 < x < 50 \\ 0, & x \ge 50 \end{cases}$$

"Unsatisfactory Known"

$$\mu_{UK}(x) = \begin{cases} \frac{x-45}{50-45}, & 45 < x < 50 \\ 1, & 50 \le x \le 65 \\ 1 - \frac{x-65}{70-65}, & 65 < x < 70 \\ 0, & x \le 45 \text{ or } x \ge 70 \end{cases}$$

"Known"

$$\mu_K(x) = \begin{cases} \frac{x-65}{70-65}, & 65 < x < 70 \\ 1, & 70 \le x \le 85 \\ 1 - \frac{x-85}{90-85}, & 85 < x < 90 \\ 0, & x \le 65 \text{ or } x \ge 90 \end{cases}$$

"Learned"

$$\mu_L(x) = \begin{cases} \frac{x-85}{90-85}, & 85 < x < 90 \\ 1, & 90 \le x \le 100 \\ 0, & x \le 85 \end{cases}$$

Fig. 1. Membership functions.

Upon Student completing a test, Fuzzy Reasoner receives Student's success rate (x) and calculates the output value for each membership function. Defaulted with easy questions for first-timers, questions difficulty level selection is eventually based upon Student's ranking during the last tests. University of Piraeus 2017 Summer School review [3] concluded that aforementioned adaptivity of test questions according to Students knowledge is both efficient and helpful. Underlying rules can be described according to the following progression sequence:

1st Test: Student is called for the first time to answer a test. Hence is automatically categorized within the fuzzy set 'Unknown' as no prior information regarding students HTML related knowledge is available. In this case, test consists of 15 easy-to-answer questions.

2nd Test: Student's knowledge level (noted as X) derives from the first test completion. As such, System produces 2nd test questions following the implementation of the bellow rule set: (a) if $X = Un$, $\mu_{Un} \epsilon [0, 1]$ or $X = UK$, $\mu_{UK} < 1$, then 15 easy-to-answer questions are selected, (b) if $X = UK$, $\mu_{UK} = 1$, then 7 easy and 8 questions of medium difficulty are selected, (c) if $X = K$, $\mu_K < 1$, then 15 questions of medium difficulty are selected, (d) if $X = K$, $\mu_K = 1$, then 7 questions of medium difficulty and 8 difficult-to-answer questions are selected, (e) if $X = L$, $\mu_L \epsilon [0, 1]$, then 15 difficult-to-answer questions are selected.

Subsequent Test: Student's knowledge level hereafter derives by means of combining the most recent test completion related level (noted as X) with the preceding test completion related level (noted as Y).

Apart from dynamically adapting test difficulty level, System stores in database Student's erroneous answers and generates proposals, recommending the review of specific material aiming to enhance Student's understanding thus facilitating Student's capacity in terms of correct answer-granting.

4 GDPR and Automated Profiling/Decision Making in ITS

4.1 Description of Profiling and Automated Decision Making Under the Scope of GDPR

The upcoming Regulation introduces: (a) milestone provisions including the Extended Personal Data Definition [16][1], the Data Protection Right in relation to Automated Decision Making and Profiling, the Data Protection Right to Object to Profiling[2] and (b) fundamental processing - related principles, Consent as a result of Lawfulness, Accountability and Transparency included. Profiling per GDPR Article 4(4) constitutes of personal data processing in order to evaluate certain characteristics relating to a person, analysing or predicting specific aspects concerning person's performance, preferences, interests, reliability, behaviour etc. and subsequently affecting decisioning in relation to aforementioned person.

Automated profiling stands as the end-product of personal data processing and subsequently is regulated by GDPR under the scope of ensuring 'the right not to be subject to a decision based solely on automated processing, including profiling.' (GDPR, Article 22). Subsequently, GDPR calls for disclosing to the data subject of 'the purposes of the processing for which the personal data are intended'. 'Meaningful information about the logic involved, as well as the consequences of such processing' is also rendered mandatory (GPDR, Articles 13–15) [17].

Initial fears over sharing to the data subject the detailed blueprints of the profiling mechanisms, have subsided by rational approaches, based on GDPR's separation of 'meaningful information' from 'significance of processing' coupled with the incorporation of the need to 'obtain an explanation of the decision reached after such assessment,' within non-binding Recital 71 rather than within binding GDPR's Articles. Recital 71 even following its April 19, 2018 amendment [18] underlines that automated processing 'should be subject to suitable safeguards, which should include specific information to the data subject and the right to obtain human intervention and to obtain an explanation of the decision reached after such assessment and to challenge the decision.' serving, in

[1] External Personal Data Definition stand for information which, either jointly with other information, or stand-alone, can identify an individual as combinedly depicted within GDPR articles 4.1 and 9.

[2] Data protection rights while re-ascertaining DPD originating ones, such as access, consent, portability and personal data profiling, map the following rights: (a) to be informed, (b) of access, (c) to rectification, (d) to erasure, (e) to restrict processing, (f) to data portability (g) to object and (h) in relation to automated decision making and profiling (GDPR articles 13–22).

the same time, as a definition of the 'right to explanation' [10]. What remains mandatory is providing data subjects with adequate insight of employed model's logic at a suitable-for-comprehension level [19]. Jointly assessed, Articles 21 and 22 suggest that the right to understand 'meaningful information' coupled by the 'significance of processing' is directly related to data subject's ability to opt out of such processing; data subject is entitled to such information related to the automated system, for an educated opt out decision to be feasible.

4.2 Case Study: Assessing ITS Paradigm Compliance with Automated Profiling/Decision Making Provisions of GDPR

Automated decision making is the driver of a dual and sequential interaction with ITS; initially assessing student's level of knowledge and consequently defining applications adaptation to student's knowledge level, propelling the most appropriate, per case, educational context and test bundle.

Student's answers are handled by the system as information for database storing and Fuzzy Reasoner feeding, the latter converting information to Student-knowledge-level-related conclusions. As such, conclusions depend on processing of answers, stored and continuously available on and from the database respectively. Conclusions drawn regarding Student's knowledge level are utilized strictly within the context of system's educational scope.

Nevertheless, system's functionality (conversion of answers granted by the student during the test session to conclusions and subsequent student profiling) is known only to system's developers and administrators. Student stands unaware of the correlation between own answers granting during the test session and knowledge level assessing, not to mention the means utilized for such a conversion. System set up breaches the **right to be informed over profiling**. Specifically, no disclaimer or note informs the Student that each time a test is answered, an AI process takes place aiming to evaluate Student's knowledge. According to the provisions of articles 12–14 GDPR the System must provide a description of AI - based conclusion extracting methodology. Description must be meaningful, utilizing simplicity and clarity, as to ensure comprehension from the simple user. In addition, per Recital 71 GDPR provisions, in order to ensure fair and transparent processing, above-mentioned description must be accommodated by a more technical and in-depth analysis, allowing for experts to validate the precision and accuracy of the process. In addition, per article 4(4) GDPR, a disclaimer stressing that data are processed, subsequently creating profiles, for educational purposes solely and that under no circumstances will sharing to third parties or using for other purposes be allowed without Student's prior and explicit consent.

Transformations, hence conclusions, originating from ITS's automated decisioning regarding students' profiling, according to students' knowledge level lead to further personal data processing, automated decisioning and adaptivity regarding student's knowledge level. According to articles 13–15 GDPR, ITS as data Controller, must provide Student with the necessary information on its functionality, emphasizing on how adaptivity, via intelligent automated decisioning, supports learning personalization [3] and guides Students through ITS's educational context.

Nevertheless, **meaningful information** in terms of logic and significance of an automated decision, per Articles 13–15 GDPR fall inconsistent to the individual decision itself, per Recital 71 GDPR. Nevertheless in our case, the mathematical mechanism employed is described along with the method used and countdown of the features selected. The fuzzy sets in verbal form along with their corresponding membership functions are stored. Upon a student completing a test, application receives from database the membership functions and feeds them with the test success rate. Verbal fuzzy sets with membership functions returning positive values are used to store Student's level of knowledge on database. As such, process involves two distinct phases, the first concerning methodology and knowledge level calculation, while the second knowledge level characterization.

Apart from Student informing upon personal data processing, system must request in a clear and precise manner **consent** for each distinct processing, for example student data storing, profile creation etc. ITS's functionalities related with profiling and automated decision making must apply only upon Student's explicit consent. However, according to article 16 GDPR, even in case of explicit consent, system should cater for Student's need of consent withdrawal, against further ITS automated decision making without any time limit or exceptions. In addition, and according to the **right to erasure or to be forgotten** (article 17 GDPR), system should support deletion of specific or all personal information assembled for the Student during Student's interaction with the application.

Student's data retaining period must be predefined. Upon duration expiration, system must request from student consensus renewal regarding personal data retaining and processing. Moreover, since system retains Student's past answers records, subsequently the right to obtain those or equally the **right of access** (article 15 GDPR) must also be materialized. In addition, Student must be able to object at any stage of the process, disputing the methodology employed.

In case of Controller evaluating that the intelligent process is indeed not valid, and its abolition or tuning is needed, then a new intelligent process should be developed, drawing from the database safeguarded answers that Students have already granted.

However advance, all information security systems risk unauthorized overriding. In order to limit the likelihood of unauthorized exploitation either of original data or data deriving from transformations and processing, ITS must **safeguard** all Student related information in cryptographic format (article 25 GDPR) and under data minimization principal, retaining only personal data that stand absolutely necessary for ITS educational (article 5 (1, c) GDPR) purposes.

Summarizing above-mentioned, self-developed ITS system incorporates elements including profiling, user personalization, artificial intelligence, automated definition of Student's knowledge level and adaptivity in terms of both educational context and tests to Student's individual needs. These elements fall under GDPR profiling and automated decision-making provisions, clashing with the right to be informed, right of access, right to be forgotten, consent and transparency.

4.3 Software Engineers Guidelines on Automated Decision Making and Profiling

In order for automated decision-making systems to achieve compliance, the following guidelines must be taken into account during ITS designing [20], categorized under the examined GDPR provisions:

Consent: (a) prior to consent, clear and adequately explanatory information on profiling is required as to ensure and enhance data subject's comprehension (b) post to profiling related consent, clear information on withdrawal is required (c) different processing purposes must follow different and distinct consent-granting (d) new processing purposes should follow consent-granting.

Right to be Informed: (a) notices must be layered and provided through-out the process. Notices to be distinct and gradual, preferably just-in-time rather than accumulative and one-off. Information overloading can be avoided by using notices with abbreviated information versions, equipped with links with full information versions (b) visualization and interaction enhances understanding.

Right of Access: (a) data used during profiling or decision-making to be categorized and documented in terms of relevance to the profiling or decision-making process (b) likewise, details of information sources, if and when used to develop profile, to be provided (c) depending on model's complexity, apart from depicting algorithms, mathematical or other appropriate forms of explanations must be provided as to allow adequate insight of the decision-making process to experts (d) data updating or amending, in an straight-forward and accessible manner, to be supported.

Meaningful information granting requires from controllers to share: (a) information used in automated decision - making process, including the source of information as well as categories of data used (b) profile-building insights, statistics and other analytics included (c) the ways by which profiling is linked to automated decision-making process affecting the data subject.

Appropriate Safeguards, specifically including: (a) coupling data minimization with minimum, from a regulatory stand point, data retention periods (b) deployment of anonymization or pseudonymization as privacy enhancing factors (c) appropriate means for data subject to challenge automated decision on a documented manner (d) certifications on processing enhanced second and third line of defense, including regular sampling reviews of the controllers' systems as to ensure not-discriminative practices and machine learning-specific auditing, as to ensure appropriate algorithms settings.

Facilitating compliance with GDPR provisions [10], above mentioned guidelines are depicted within bellow use case diagram in UML (Fig. 2).

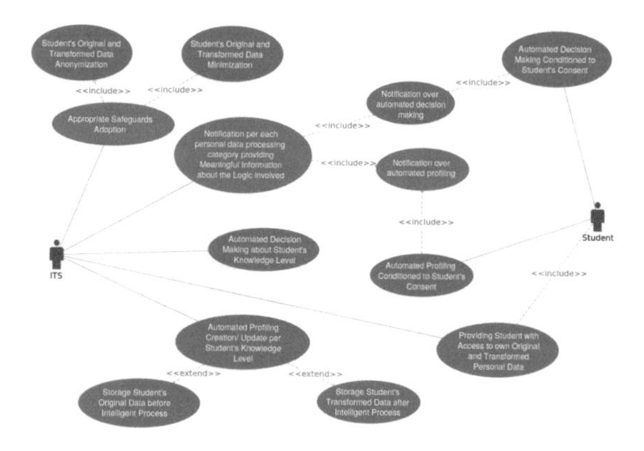

Fig. 2. Guidelines use case diagram.

5 Conclusions

GDPR imminent enforcement calls for organized cross-functional cooperation, IT and Legal Science prevailing. Cooperation aims in ensuring compliance of existing as well as new application by amending or adding functionalities. But most importantly, IT and Legal Science cooperation will offspring new software engineering standards, facilitating the development of new applications encompassing GDPR's personal data protection by design.

Despite significant steps towards this direction, by the time of writing, no concluded research on GDPR compliance, targeted to ITS was available. ITS are systems of increasing acceptance, due to their effectiveness as educational means. Authors focused in assessing and addressing potential compliance gaps using a Fuzzy Logic based ITS paradigm.

System's intelligence is based upon Fuzzy Logic theory, in order to define Students' knowledge level and subsequently questions difficulty level, in other words producing conclusions as a result of fuzzy reasoning. Conflict areas of ITS with GDPR where pin-pointed, predominantly regarding automated profiling or personalization and automated decision making or adaptivity and secondary regarding right to be informed, user consent, right of access and right to be forgotten; the secondary category always under the light of System's personalization and adaptivity.

Under GDPR scope, research emphasized on directly inputted personal data (for example student's registration data), as well as interaction-deriving-data (for example answers to tests), conversion of data to conclusions followed by conclusion's utilization in terms of automated profiling (or personalization under ITS scope) and automated decision making (or adaptivity under ITS scope).

References

1. Zhang, B., Jia, J.: Evaluating an intelligent tutoring system for personalized math teaching. In: 2017 International Symposium on Educational Technology (ISET), pp. 126–130. IEEE (2017)
2. Göbel, S., Wendel, V.: Personalization and adaptation. In: Serious Games, pp. 161–210. Springer (2016)
3. Papadimitriou, S., Virvou, M.: Adaptivity in scenarios in an educational adventure game. In: 2017 8th International Conference on Information, Intelligence, Systems & Applications (IISA), pp. 1–6. IEEE (2017)
4. Streicher, A., Smeddinck, J.D.: Personalized and adaptive serious games. In: Entertainment Computing and Serious Games, pp. 332–377. Springer, Cham (2016)
5. General Data Protection Regulation (GDPR): Regulation (EU) 2016/679 of the European parliament and of the council of 27 april 2016 on the protection of natural persons with regard to the processing of personal data and on the free movement of such data, and repealing directive 95/46. Off. J. Eur. Union (OJ) **59**, 1–88 (2016)
6. Mougiakou, E., Virvou, M.: Based on GDPR privacy in UML: Case of e-learning program. In: Proceedings of 2017 8th International Conference on Information, Intelligence, Systems & Applications (IISA), pp. 1–8. IEEE (2017)
7. Directive 95/46/EC of the European parliament and of the council of 24 october 1995 on the protection of individuals with regard to the processing of personal data and on the free movement of such data. Off. J. EC 23(6), 31–50 (1995)
8. Castelluccia, C., Druschel, P., Hübner, S., Pasic, A., Preneel, B., Tschofenig, H.: Privacy, accountability and trust-challenges and opportunities ENISA (2011). http://www.enisa.europa.eu/activities/identity-and-trust/library/deliverables/pat-study/atdownload/fullReport
9. Layton, R.: How the GDPR compares to best practices for privacy, accountability and trust (2017). https://ssrn.com/abstract=2944358
10. Wachter, S., Mittelstadt, B., Floridi, L.: Why a right to explanation of automated decision-making does not exist in the general data protection regulation. Int. Data Priv. Law **7**(2), 76–99 (2017)
11. Goodman, B., Flaxman, S.: Eu regulations on algorithmic decision-making and a right to explanation. In: ICML Workshop on Human Interpretability in Machine Learning (WHI 2016), New York (2016). http://arxiv.org/abs/1606.08813v1
12. Chrysafiadi, K., Virvou, M.: Fuzzy logic for adaptive instruction in an e-learning environment for computer programming. IEEE Trans. Fuzzy Syst. **23**(1), 164–177 (2015)
13. Zapata-Rivera, D.: Adaptive, assessment-based educational games. In: International Conference on Intelligent Tutoring Systems, pp. 435–437. Springer, Heidelberg (2010)
14. Ross, T.J.: Fuzzy Logic with Engineering Applications. Wiley, New York (2009)
15. Chrysafiadi, K., Virvou, M.: Evaluating the integration of fuzzy logic into the student model of a web-based learning environment. Expert Syst. Appl. **39**(18), 13127–13134 (2012)
16. Borking, J.: Der identity protector. Datenschutz und Datensicherheit **20**(11), 654–658 (1996)
17. Kingston, J.: Using artificial intelligence to support compliance with the general data protection regulation. Artif. Intell. Law **25**(4), 429–443 (2017)

18. Council of European Union: Corrigendum deed amending several provisions of GDPR (2018) (Interinstitutional File: 2012/0011 (COD) 8088/2018), 19 April 2018
19. Burt, A.: Is there a right to explanation for machine learning in the GDPR? IAPP, 1 Jun 2017
20. Party, A.D.W.: Article 29 data protection working party guidelines on automated individual decision-making and profiling for the purposes of regulation 2016/679 2017, October 2017

Authoring Technological and Platform Independent Learning Material and Student's Progress Profile Using Web Services

Spyros Papadimitriou(✉), Konstantina Chrysafiadi, and Maria Virvou

University of Piraeus, Karaoli & Dimitriou Street 80, 18534 Piraeus, Greece
{spap,kchrysafiadi,mvirvou}@unipi.gr
http://www.cs.unipi.gr

Abstract. Nowadays the rapid development of computer and Internet technologies in the field of education has changed the ways of teaching and learning. The production of adaptive educational applications and systems that enrich the tutoring and learning processes with "intelligence", in order to adapt either the learning material or the tutoring and learning processes to each individual student's needs and abilities, offering her/him a personalized learning experience. However, building an adaptive educational environment and/or application is difficult and complex, since either technical knowledge and programming skills or the expertise of tutors are demanded. In this paper an authoring tool, which offers the instructors the possibility to create learning material and student's profile that are technological and platform independent, is presented. The innovative operation of the presented authoring tool is based on web services and is due to the following facts: (i) the created learning content can be used by any educational application regardless of the system's technological characteristics or programming language, (ii) the created progress profile of a student, who uses two different educational applications that call the same web services of the authoring tool, is recognized by both applications. Therefore, if the knowledge level of a student is changing during her/his interaction with the one application, then the other application will, also, recognize the updated knowledge level of the particular student.

Keywords: Authoring tools · Web services
Educational applications · Educational games · Student profiling

1 Introduction

Towards the past decade the rapid development of computer and Internet technologies in the field of education has changed the ways of teaching and learning. Digital education and e-learning applications offer easy access to knowledge domains and learning processes from everywhere for everybody at any time.

M. Virvou et al. (Eds.): JCKBSE 2018, SIST 108, pp. 242–251, 2019.
https://doi.org/10.1007/978-3-319-97679-2_25

Furthermore, information and computer technologies are considered some of the most beneficial teaching tools supporting student learning in the classroom [1]. The related scientific literature deals with the incorporation of "intelligence" into the teaching and learning processes of digital educational applications. Particularly, there is an increased interest in adaptive educational systems that recognize each individual learner's needs, misconceptions and abilities and provide her/him with a personalized learning experience. However, building an adaptive educational environment and/or application is difficult and complex, since either technical knowledge and programming skills or the expertise of tutors are demanded. These drawbacks can be resolved by the use of the authoring tools, which aim at providing environments for cost-effective development of tutoring systems that can be intelligent and adaptive to individual students [2].

The main purpose of an authoring tool is to author the knowledge domain model, which stores the learning material that is taught to students. Therefore, many tutors should use an authoring tool in order to author the course modules of the subject-matter domain, which involve domain concepts, exercises, quizzes and other learning material. An authoring tool has to allow teachers to create their own educational material for the topic they desire, without experiencing technology as an obstacle in the authoring process. Furthermore, the created learning material should be reused, modified and shared in an effortless way [3]. Many authoring tools of adaptive educational systems attempt to give the instructors the possibility of configuring different aspects of the student model [4–6]. The aim is to give the instructors the possibility to monitor the learners' performance and progress in order to provide them personalized advise and learning support.

However, the learning content that is created through the most authoring tools deals with technological limitations that do not allow it to be reused in different platforms or educational systems. Furthermore, in many circumstances the learning content, which has been authored through an authoring tool, has to be configured taking into consideration many parameters in order to can be used in a third application or system. Therefore, the challenge is to create an authoring tool that gives the possibility to create learning content that is either technological or platform independent. Web services is a solution for this challenge. A Web service is a collection of functions that operate as a single entity and are available through the network for use by other applications. They are frequently used in modern software engineering and application development due to their numerous advantages, including interoperability, reusability, composability and near global accessibility via the Web [7]. They represent a form to render distributed resources connected to the Internet. Using web services to offer functionality through the web has become more common each day due to its flexibility and security when compared with other ways to offer functionalities and because of its scalability [8].

In view of the above, in this paper we present an authoring tool that offers the instructors the possibility to create learning material and student's profile are technological and platform independent. The particular authoring tool gives the tutor the possibility to author: (i) the desired knowledge domain in a hierarchical structure that includes courses, chapter, lessons, etc., (ii) exercises and quizzes for the assessment of students, (iii) the student's profile that includes her/his knowledge level and performance in order to be able to monitor her/his progress. The innovation of the presented authoring tool is due to the fact that the created learning content can be used by any educational application (web, mobile, game etc.), regardless of the system's technological characteristics or programming language. This is realized by using web services. Furthermore, the profile of a student, who uses two different educational applications that call the same web services of the authoring tool, is the same for both applications. This means that if the student's knowledge level is changing through one application, the other application will know it.

2 Related Work

In literature review there are many authoring tools that allow instructors to create and edit learning activities and material, student models or adaptive tutoring systems. For example, EDUCA is a software tool for creating adaptive learning material in a Web 2.0 collaborative learning environment [9]. The material is initially created by a tutor/instructor and later maintained and updated by the user/learner community to each individual course. Also, Virvou and Troussas have presented a knowledge-based authoring tool for authoring multilingual systems that can create individual and personalized student models for each student [4]. E-Adventure is another authoring tool, developed by Torrente et al. which is not only focused on the abstraction of the programming tasks, leaving the technical features of the game to programmers and media designers, but also on supporting those aspects that are specific of the educational domain, usually not contemplated in toolkits for commercial videogames [10]. In addition, Roldán-Álvarez et al. have created DEDOS-Editor, which allow teachers to design their own learning activities, and DEDOS-Web, which allows the students to perform those activities adapting them to multiple devices [3]. Furthermore, Llinás et al. have presented a set of teacher authoring tools to facilitate the tasks of identifying learning tricky topics, noting student difficulties, and creating quizzes for knowledge evaluation [6]. Another authoring tool is ViSH Editor, an innovative web-based e-Learning authoring tool that aims to allow teachers to create new learning objects using e-Infrastructure resources [11].

However, after a thorough investigation in the related scientific literature, we came up with the result that there was no implementation of authoring tools, which can be used for creating either learning material that can be imported in any digital educational system, or student's profile that can be recognized and updated by different educational applications that are used by the particular student. In other words, the gain of the presented authoring tool is that the created learning material and student's profile are technological and platform independent. This is realized by using web services. The created learning material can be used by any educational application (web, mobile, game etc.), regardless of the system's technological characteristics or programming language. In addition, a particular student's profile and progress is recognized by any application that is used by the particular student and is calling the same web services that have been created through the presented authoring tool.

3 System Architecture

The system consists of two distinct components, the authoring tool and the web services component (Fig. 1). The first is used by the tutor to create learning material and monitor the performance of her/his students. It also includes the quiz question selection logic and the student's knowledge level assessment methodology. These functionalities are provided as services to any external application that wants to exploit them, regardless of the implementation platform.

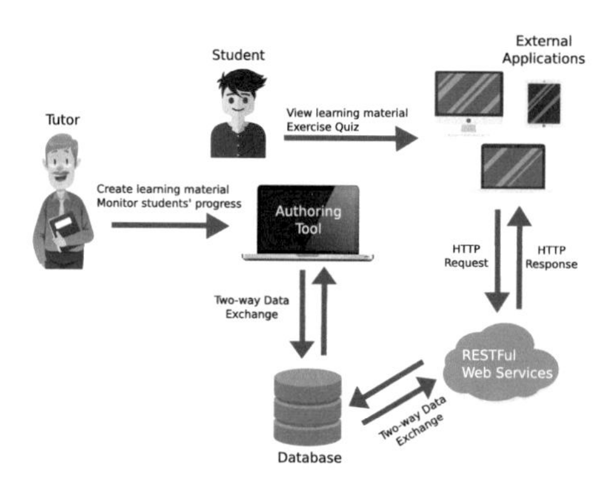

Fig. 1. System architecture.

3.1 Authoring Tool Description

An authoring tool enables teachers who have no specialized knowledge about technology to easily create their own learning material and monitor their progress of their students through quizzes Our tool is designed with simplicity in mind so that it is fully understandable and clear for tutor to use (Fig. 2). The educational content that may be inserted follows the same hierarchical structure for all tutors (Fig. 3). Categories are entities where courses can be integrated, for example "Computer Science", "High School Courses" or "Language Learning". Categories can be divided into subcategories. For example, the category "Computer Science" can be further divided into "Software Development" and "Algorithms" subcategories while the category "Language Learning" can be divided into "Learning English" and "Learning Greek". A course can be associated with multiple categories and subcategories. Thus, course "Introduction to HTML5" can be associated with category "Computer Science" and its subcategory "Web Development". Each course is divided into individual chapters and each chapter

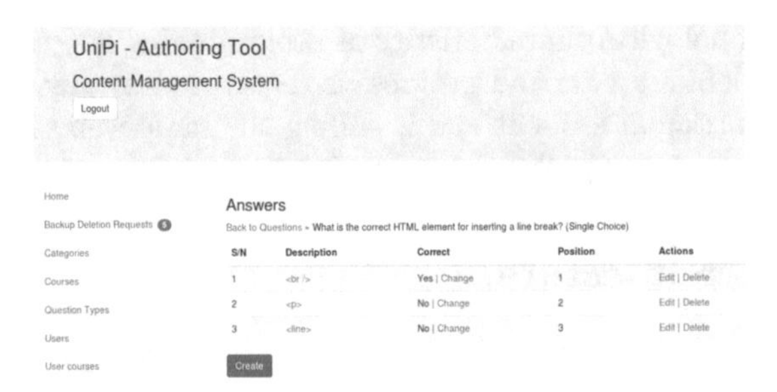

Fig. 2. Authoring tool interface.

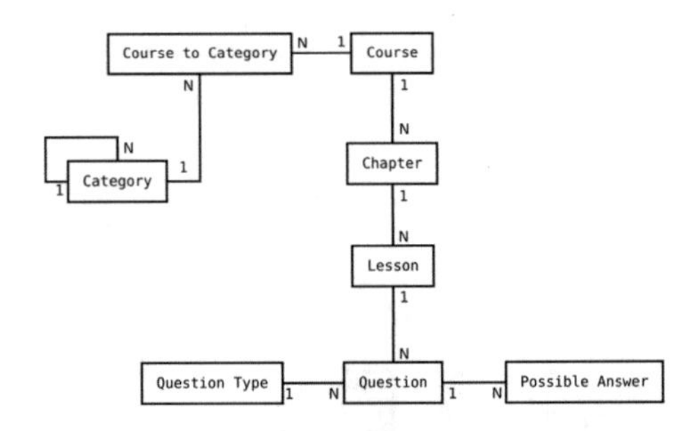

Fig. 3. Learning material hierarchy.

consists of many lessons. For each lesson the teacher can create questions and possible answers. A question may belong to one of the following types: single choice (e.g. true or false), multiple choice, fill-the-gaps and sorting. In addition, each question relates to a specific level of difficulty: easy, medium, difficult. The number of possible answers to each question has no limitation as the teacher can create as many as s/he desires. A paradigm of the hierarchy of learning content is depicted below:

- Computer Science
  - Web Development
    - Building Web Applications in PHP
      - HTML
        - HTML tags
        - Headings
          - Choose the correct HTML element for the largest heading.
            - <h6>
            - <heading>
            - <h1>
      - CSS
        - ...
      - PHP
      - ...

The teacher has access only to the courses s/he has created and can only monitor the performance of students who have enrolled in these courses. Furthermore, s/he can view all the quizzes their students have taken, when they were created, if and when they were completed and how they were answered. The student can exercise in a lesson, chapter or the entire course. When the student completes a quiz, the system silently performs an intelligent process that evaluates and characterizes the student's knowledge level in terms of lesson, chapter or course. The system can give the following characteristics regarding the student's level of knowledge: (i) unknown, (ii) unsatisfactory known, (iii) known, and (iv) learned. In this way the system represents how well a particular student knows the educational content. It is noted that the levels of difficulty of the questions and the possible values of the student's level of knowledge are defined by the administrator of the entire authoring tool and not by the teachers. In this way, we achieve the standardization of the structure of learning content and knowledge levels in order to avoid the possible confusion of the student choosing to study courses belonging to different teachers.

3.2 Interoperability - Web Services Component

Our authoring tool allows the teacher to create content for any educational subject without any limitation. However, the main originality of our authoring tool is to allow an external application, regardless of its development technology, to use the authoring tool's educational content. The external application can be developed in any language, such as PHP, Java, C#, and deployed to any platform, e.g. desktop, web or mobile. The communication between our system and the third application is achieved by using web services based on the architectural style of Representational State Transfer (REST) [12]. The third system can send HTTP requests to the appropriate Uniform Resource Identifiers (URIs) of our system and receive authoring tool's objects in Javascript Object Notation (JSON). The choice of JSON format was based on the fact that it is lightweight data-interchange format and can be easily generated or parsed by a machine [13]. The system supports the following functionalities through web services:

- Register student.
- Authenticate student.
- Retrieve learning content (course, chapter or lesson).
- Create a quiz about a course, chapter or lesson.
- Retrieve current question of a quiz.
- Submit answer regarding current a quiz question.
- Retrieve the student's knowledge regarding a course, chapter or lesson.

Table 1 illustrates two of the above-mentioned functionalities and their corresponding JSON response examples. If the student uses two or more applications that call the services of the authoring tool, the profile is affected by all applications. The latter have access to the same student's profile and the same information that accompanies it. If the student completes a quiz through one application

Table 1. Web services methods examples

Method description	HTTP request - URI	JSON response example
- Student registration	/user/register	{code":200,"message": "Account is created successfully."}
- Retrieve the student's knowledge regarding a course, chapter or lesson	chapter/model/14	{id_user":"5","id_chapter":"14", "knowledge_level":"learned"}

and changes her/his level of knowledge, the other applications will know about this change. Additionally, if the student starts a lesson quiz in one application, the same questions will appear in other application until the student completes the quiz in any of the external applications. This is due to the fact that there is only one active quiz per lesson, chapter and course.

4 Examples of Use

Below we describe two examples of different educational applications (an educational web Flash game and an educational PHP web application) to illustrate the operation of the presented authoring tool.

4.1 Educational Adventure Game

The HTML Escape Game is an educational adventure game that is used for learning HTML [14]. In that game, the player's main objective is to help the hero to escape from the building in which he is imprisoned (Fig. 4).

The game calls the web services relating to user registration/authentication, creating HTML-related quizzes, submitting quiz answers, and retrieving the player's knowledge level after completing a quiz. Based on the latter, the game decides whether and how it will reward the player. The game has no involvement

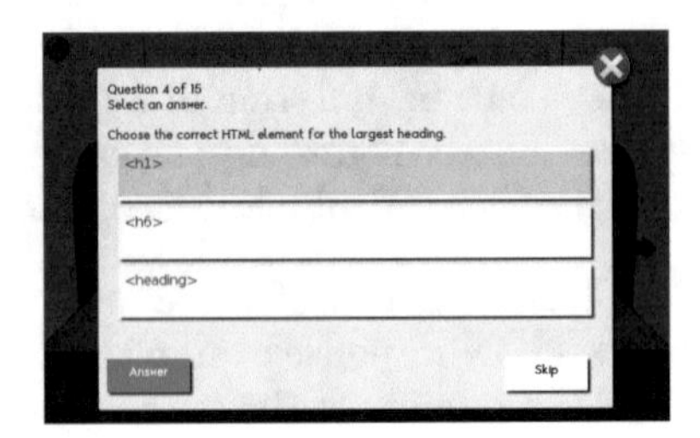

Fig. 4. Educational adventure game based on authoring tool.

in selecting HTML questions or calculating the player's knowledge as these have been introduced through the authoring tool and are part of the business logic of the web services that accompany it. Therefore, anything related to the educational part of the application is performed by our system without the game being interested or aware of the way this is done. The game just retrieves JSON objects and chooses how to handle them, e.g. decides how to display a quiz or executes specific actions depending on the student's knowledge level after completing a quiz (Fig. 4).

4.2 Web-Based Educational Application

An original integrated e-training environment for teaching programming concepts and structures through pseudo-code have been developed [15]. It has been used by the students of a postgraduate program in Informatics, at the University of Piraeus. The particular educational application offers personalized e-training in programming. The learning material of the particular application is broken down into several domain concepts: declarations of variables and constants; expressions and operators; input and output expressions; the sequential execution of a program; selection and iteration programming structures; arrays; and functions. These domain concepts that constitute the learning material are organized into chapters and lessons. They have been created using the authoring tool that is described above. An example of the created learning material is presented in Fig. 5. Furthermore, the particular web-based educational application offers personalized tests that include four types of exercises: (1) true/false exercises; (2) multiple choice exercises; (3) fill in the gap space exercises, where the student fills in a symbol or a command in order to complete a program; and (4) exercises in which users put certain parts of a program in the right order. Either the questions or their answers have been created by using the presented authoring tool.

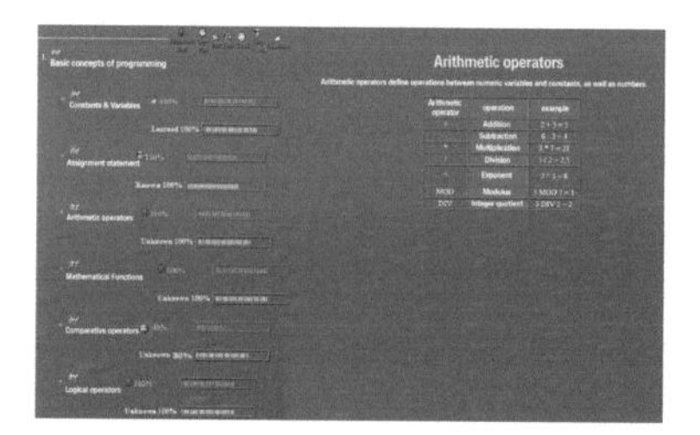

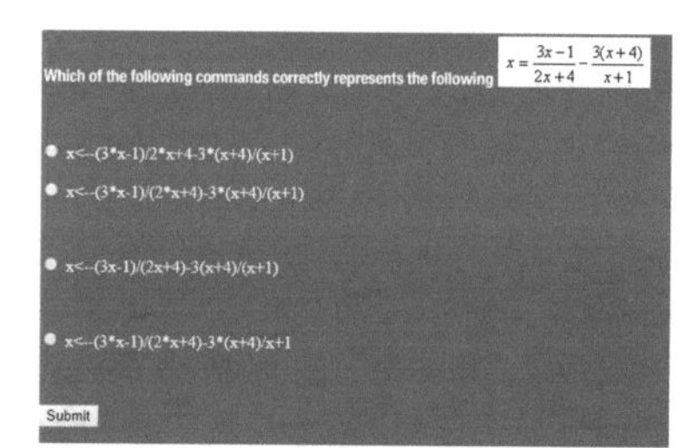

Fig. 5. Web-based educational application based on authoring tool.

4.3 Common User Profile

The HTML Escape game used the same web services of the authoring tool with the web-based educational application that is described above. The only difference was the learning object used by each application. In particular, the HTML escape game called the web services that allowed the game to retrieve the quizzes about "HTML programming" course while the web-based application retrieved the content about "Computer Programming". However, we adjusted the web service input parameters so that the game can be used for teaching the same course with the web-based application. We have noticed that a student, who used both applications, was recognized either by the educational game or by the web-based educational application. In more details, the particular student played the game solving quizzes that correspond to the learning material of "Computer Programming". While, s/he was playing, her/his knowledge level changed. When, s/he logged in the web-based educational application, her/his knowledge level changes were recognised and the educational application adapted the learning material to the new knowledge level of the particular student.

5 Conclusions

In our article, we presented an authoring tool that uses a simple and clear hierarchical structure that hosts the tutor's educational material. In addition, the tool incorporates intelligent procedures to create the student's profile and produce exercises tailored to her/his knowledge level.

Its main innovation is that it allows external educational applications to make use of its capabilities and the content of the teachers. This is achieved through the use of REST Web Services and the exchange of information in JSON format. The external application is exempt from the business logic of the educational part and focuses only on the calls of the web services and the management of their responses. The way in which a chapter or a quiz will be displayed, or the choice of actions to be performed based on the student's knowledge and performance, are determined exclusively in the context of the external application. The implementation technology is independent of the tool's technology and the final application can be deployed on any platform, desktop, web and/or mobile. The learning material for teachers and student's profiles and performance are saved on a single point, accessible from any external application that consumes the web services. Each application has access to the same data and any changes to them are visible in all applications.

System improvements and new features integration will be accessible and available for exploitation by all external applications. In view of this, our future plans include upgrading the tool and adding new characteristics that will offer even better personalized user experience. Furthermore, we are implementing new applications, such as 3d games and educational mobile applications, which will utilize the authoring tool and will be used for learning any course registered in it.

Acknowledgments. This work has been partly supported by the University of Piraeus Research Center.

References

1. Lamb, R.L., Annetta, L., Firestone, J., Etopio, E.: A meta-analysis with examination of moderators of student cognition, affect, and learning outcomes while using serious educational games, serious games, and simulations. Comput. Hum. Behav. **80**, 158–167 (2018)
2. Virvou, M., Alepis, E.: Mobile educational features in authoring tools for personalised tutoring. Comput. Educ. **44**(1), 53–68 (2005)
3. Roldán-Alvarez, D., Martín, E., Haya, P.A., García-Herranz, M., Rodríguez-González, M.: DEDOS: an authoring toolkit to create educational multimedia activities for multiple devices. IEEE Trans. Learn. Technol. (2018)
4. Virvou, M., Troussas, C.: Knowledge-based authoring tool for tutoring multiple languages. In: Intelligent Interactive Multimedia Systems and Services, pp. 163–175. Springer (2011)
5. Moundridou, M., Virvou, M.: Analysis and design of a web-based authoring tool generating intelligent tutoring systems. Comput. Educ. **40**(2), 157–181 (2003)
6. Llinás, P., Martín, E., Hernán-Losada, I., Gutiérrez, M.A., Clough, G., Adams, A.: Authoring tools supporting novice teachers identifying student problems. In: Design for Teaching and Learning in a Networked World, pp. 592–595. Springer (2015)
7. Syu, Y., Kuo, J.Y., Fanjiang, Y.Y.: Time series forecasting for dynamic quality of web services: an empirical study. J. Syst. Softw. **134**, 279–303 (2017)
8. Garcia, C.M., Abilio, R.: Systems integration using web services, rest and soap: a practical report. Revista de Sistemas de Informação da FSMA **19**(19), 34–41 (2017)
9. Cabada, R.Z., Estrada, M.L.B., García, C.A.R.: EDUCA: a web 2.0 authoring tool for developing adaptive and intelligent tutoring systems using a kohonen network. Expert Syst. Appl. **38**(8), 9522–9529 (2011)
10. Torrente, J., Del Blanco, Á., Marchiori, E.J., Moreno-Ger, P., Fernández-Manjón, B.: <e-adventure>: introducing educational games in the learning process. In: Education Engineering (EDUCON), 2010 IEEE, pp. 1121–1126. IEEE (2010)
11. Gordillo, A., Barra, E., Gallego, D., Quemada, J.: An online e-learning authoring tool to create interactive multi-device learning objects using e-infrastructure resources. In: Frontiers in Education Conference, 2013 IEEE, pp. 1914–1920. IEEE (2013)
12. Garriga, M., Mateos, C., Flores, A., Cechich, A., Zunino, A.: Restful service composition at a glance: a survey. J. Netw. Comput. Appl. **60**, 32–53 (2016)
13. Crockford, D.: The application/JSON media type for javascript object notation (JSON) (2006)
14. Papadimitriou, S., Virvou, M.: Adaptivity in scenarios in an educational adventure game. In: 2017 8th International Conference on Information, Intelligence, Systems & Applications (IISA), pp. 1–6. IEEE (2017)
15. Chrysafiadi, K., Virvou, M.: Fuzzy logic for adaptive instruction in an e-learning environment for computer programming. IEEE Trans. Fuzzy Syst. **23**(1), 164–177 (2015)

Computerized Adaptive Assessment Using Accumulative Learning Activities Based on Revised Bloom's Taxonomy

Akrivi Krouska, Christos Troussas, and Maria Virvou(✉)

Software Engineering Laboratory, Department of Informatics, University of Piraeus, 18534 Piraeus, Greece
{akrouska,ctrouss,mvirvou}@unipi.gr

Abstract. The need of developing more efficient educational systems leads to the incorporation of personalized operations. Digital learning focuses mainly on the adaptive content and navigation. However, the provision of a valid assessment tool is essential to an integrated e-learning system, as it indicates the accomplishment of learning goals. One such testing model should be designed based on the new demands to knowledge, skills and attitudes that students have to acquire, and considering the students' needs. To this direction, this paper introduces an adaptive assessment system where the test items are designed based on Revised Bloom's Taxonomy and the assessment content is adapted to students. One major advantage of the proposed system is that it provides a better detection of student's learning gaps, useful for further system adaptivity.

Keywords: Adaptive testing · E-Learning · Revised Bloom's Taxonomy Cognitive development · Accumulative learning activities

1 Introduction

The proliferation of Information and Communications Technology (ICT) brings new technological challenges in educational field. During the earliest years of its emergence, computers just acted as a tutor providing "one size fits all" computer assisted instruction [1]. Nowadays, the advent of Internet and the incorporation of Artificial Intelligence techniques in educational software lead to the development of e-learning systems adaptive to learner's needs. Thus, the trend of digital learning is to provide personalized instruction based on teaching strategies and student's profile, focusing mainly on adapting content presentation and navigation [2].

The assessment process is a significant issue in a curriculum. Through this process, students' weaknesses are identified, their knowledge is measured and their skill acquisition is defined [3]. The outcomes of students' testing are used to provide personalized services to them and adapt the tutoring system to their individual needs. Moreover, these outcomes are useful to analyze the achievements of learning process, and thus, to evaluate the quality of instructional systems. In the view of above, the design of a proper assessment tool is essential for developing integrated e-learning environments.

M. Virvou et al. (Eds.): JCKBSE 2018, SIST 108, pp. 252–258, 2019.
https://doi.org/10.1007/978-3-319-97679-2_26

Many e-learning systems provide a simple testing tool, which is fixed to all students and evaluates only the recall of learning material's meaning. However, an effective instructional environment should incorporate a valid and efficient assessment model. One such proper assessment adopts a pedagogical theory, evaluates the acquisition of high level cognition, namely meaningful/associative/active learning in addition to rote learning, and adapts its question items to help students reach the best learning outcomes for them and the objectives of the curriculum [4].

This paper presents an adaptive assessment module, developed in the context of an integrated e-learning platform, which accumulates learning activities based on Revised Bloom's Taxonomy (RBT) and tailors them to student's profile. The RBT provides a framework for classifying the learning objectives into the different levels of cognition [5]. This taxonomy is useful for course and assessment design, as it can lead students from the simplest cognition types, remembering and understanding, to the more complex ones, evaluating and creating. Moreover, assessment's adaptivity to student's training needs enhances the reliability of learning outcomes and offers a better diagnosis of student's weaknesses, useful for the personalized instruction and the recommendation of learning objects.

The rest of paper structures as follows. In Sect. 2, the literature review of research area is reported. Section 3 refers to the innovative assessment tool developed. In Sect. 4, some case studies to understand the system operation are described. Finally, conclusions and future work are presented.

2 Literature Review

Defining the learning objectives is essential for the success of an instructional process and refers to the desired learning outcomes that are expected to be achieved in the end of the curriculum. These objectives should be defined by the instructors clearly and meet students' needs. The way in which the learning objectives are specified is very important in order to be comprehensible to the students what it is anticipated to learn after the course complication, the learning goals they have to achieve and what it is needed to complete the course successfully.

Revised Bloom's Taxonomy (RBT) is a well-known framework providing a classification of learning objectives in three major domains: the cognitive, the affective and the psychomotor one [5]. This paper focuses on the cognitive domain, where the objectives are related to the acquisition of knowledge and the development of skills. Cognitive objectives vary from simple recall of definitions to high order cognitive skills, like analyzing and synthesizing new ideas. The taxonomy is divided into six levels: *Remembering, Understanding, Applying, Analyzing, Evaluating* and *Creating*. Moreover, it is hierarchical; meaning that each level of cognitive development is dependent on the attainment of previous level's goals. Figure 1 illustrates the RBT levels.

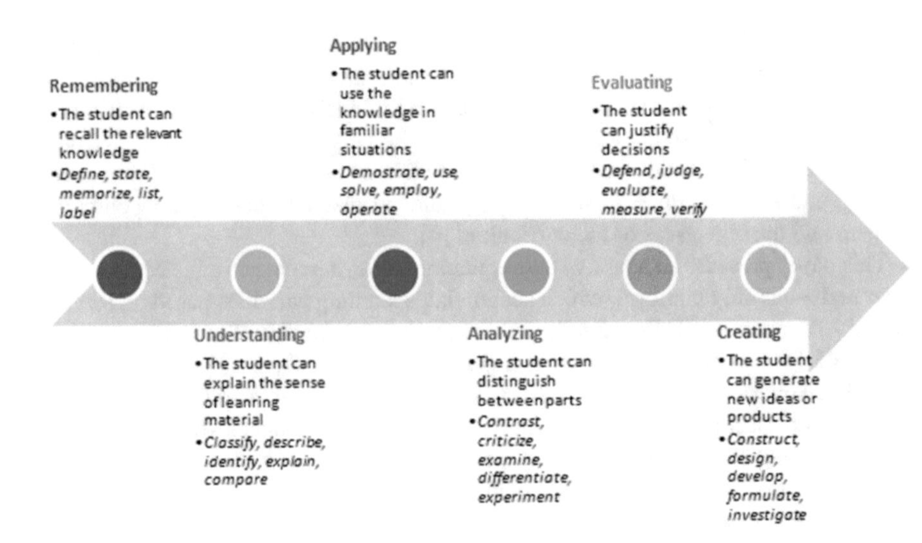

Fig. 1. Revised Bloom's Taxonomy

RBT can be helpful for specifying correctly course objectives, preparing valid and efficient assessments by including activities related to different taxonomy's levels and improving students' cognitive level. Many researchers have applied RBT to the development of course testing system. In [6], the authors proposed a web-based programming assisted system for supporting programming activities with various difficulty levels of cognition based on Bloom's taxonomy. In [7], the research deals with a workflow modeling of tests where question items follow the knowledge dimensions of RBT and are chosen according the logical and meaningful connections between them and learners' results on them. In [8], the authors propose a computer-based assessment of problem solving using the validity of RBT. In [9], the paper describes the creation of e-learning exercises according to RBT and compares its effect with the success rate of pupils' achievement. In [10], the research focuses on the implementation of a diagnosis e-assessment model that uses RBT in order to formulate the behavioral objectives to be measured.

3 Computerized Adaptive Assessment to RBT

RBT proposes a classification for writing learning objectives, which has been applied widely in education. According to this taxonomy, the cognitive domain divided into six levels of cognitive development. When designing RBT-based assessment system, the test assignments should apply and use a combination of these levels in order to evaluate intensively the student's acquisition of knowledge and skills. Moreover, the system should adapt its assessments depending on student's response and profile.

Figure 2 illustrates the framework for developing adaptive RBT-based assessment system. Firstly, the instructor designs the course following RBT. This means that the instructor needs to define the learning objectives and declare which cognitive developments are evaluated in each course test. Moreover, he/she has to create the corresponding question items of the assessments for each cognition level that the tests examine. Afterwards, these items are implemented using the system's authoring tool. The system provides a variety of question types in order the instructor to choose the proper one for each level. Finally, students explore the e-learning system and when they ask to test their knowledge, the system provides them an adaptive RBT-based assessment.

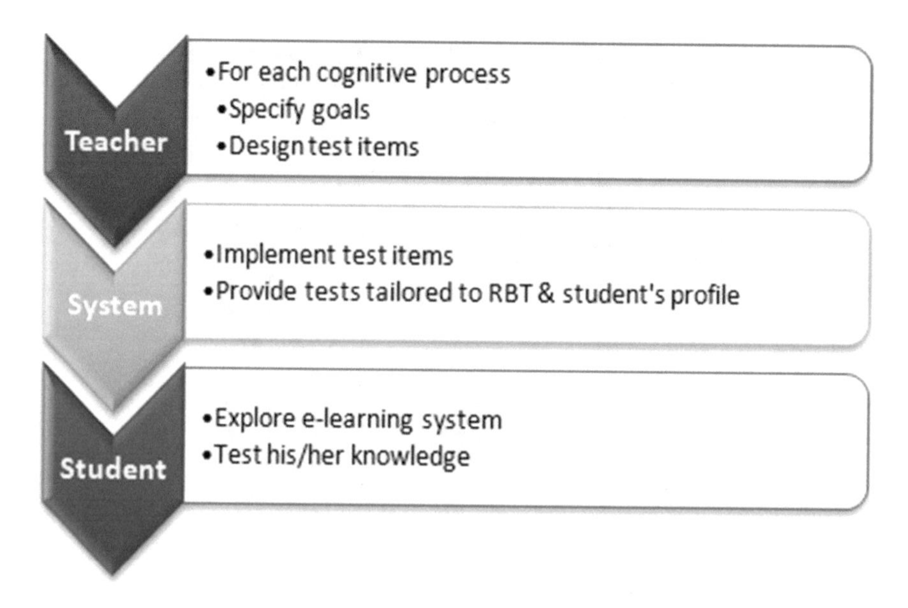

Fig. 2. Adaptive RBT-based assessment tool framework

Given that the structure of the assessments follows RBT levels, it examines hierarchically the cognitive processes defined by the instructor, through a set of questions derived from an accumulative learning activities repository. In order to move to higher cognition level it is necessary to acquire the knowledge and skills of the previous level. The test starts with questions from lower order cognition, and if the required knowledge and behavior are acquired, questions from next level are displayed. Thus, the test content is adaptive and depends on the student responses and profile. In particular, the system firstly generates three question items of the lower level knowledge adjusted their difficulty to student profile from previous test attempts. For instance, if the student has low achievements in tests, the system selects easy to intermediate difficulty questions, while if the student is advanced, high difficulty questions are chosen. The selection of next question item depends on student responses. If there are wrong answers, an error diagnosis process is applied. Based on this error diagnosis, it is decided if the current cognition level would be examined further displaying more questions on this level, or if the test would continue to the next level, or if the evaluation would terminate as a learning gap is detected. If there are no wrong answers, the test continues to the next cognition level generated question items in the same way.

The system provides an authoring tutor giving the opportunity to teacher to design tests with different question types depending on RBT levels [11]. Exercises like true-false, multiple-choice and fill-in-the-blank, are appropriate for remembering, understanding, applying and analyzing level. Moreover, matching and short-answer exercises can be used in these levels. Regarding evaluating and creating level, the more proper exercises are the open questions, as it is examined the critical thought of students and the capability to produce new facts based on the acquired knowledge. However, it can also be used closed type questions, such as multiple-choice, that are designed in a way that the student has to choose the correct answer that emerges from the review of data or corresponds to new models and assumptions. Table 1 presents the appropriate question types for each cognitive development.

Table 1. Question types corresponding to RBT levels

RBT level	What tests	How to test
Remembering	The recall of definitions	Multiple choice True/False Fill-in-the-blank Matching Short-answer
Understanding	The understanding of the sense of concepts	Multiple choice True/False Matching Crossword Short-answer
Applying	The capability of applying the knowledge to new situations	Multiple choice Puzzle Practice exercise Problem-solving
Analyzing	The analysis of concepts into and their relationships	Multiple choice Rank order Matching Concept maps Open question
Evaluating	The capability of making judgments	Multiple choice Open question
Creating	The capability of producing new facts	Multiple choice Essay

The proposed evaluation model has the following advantages: a. the student's weaknesses are detected easily, b. the difficulty of question items is increased stepwise, c. entire course material is examined, d. the learning process is better organized, and e. new student skills are developed.

4 Cases of Operation

In this section, examples of operation are presented. The course that is implemented is the Biology. The course consists of five chapters and a test is aligned to each chapter. The tests evaluate certain cognitive development and the number of questions and the questions' content are dynamically defined depending on student profile and responses to test.

David is a new student in the e-learning system and started studying Chapter 1. In the end of chapter, he tested his knowledge. Test 1 examines the lowest four cognitive developments. As it was his first test, the system had no information about his level of knowledge and learning needs for the selection of the first set of questions. Therefore, it was chosen three staggering difficulty questions for remembering level. He answered correctly to all of them and the system selected the next three questions for understanding level. David made a mistake to the intermediate difficulty question which the error diagnosis module characterized it as probably incautious mistake. Hence, the system generated the next level questions. In this level, David made one mistake that showed that he probably had a gap in applying the knowledge. So the system chose one more question for this level. David did again a serious mistake, leading the system not to examine the next level, the analyzing, as a learning gap is detected. Thus, the test completed, providing advice concerning the error diagnosis and adapting the learning process properly.

Helene is an excellent student, having passed the tests of three first chapters successfully without mistakes. Now she tested in Chapter 4. This test evaluates all RBT levels. Regarding her profile, the system generated high difficulty questions. In evaluating level, she made a mistake which characterized by the error diagnosis as probably learning gap. Thus, one more question on this level was generated, intermediate difficulty this time. Helene answered correctly and for the next level the system adapted its selection of questions to intermediate/high difficulty. Helene answered without mistake and the test completed giving her its feedback.

5 Conclusions

The evolution of Information and Communications Technology (ICT) introduced new technological advances in education, enhancing the learning process provided by e-learning systems. One of the innovations is the present assessment tool which combines a pedagogical theory with new technologies. In particular, the testing process uses accumulative question items based on RBT and adapts them in order the students to achieve the best learning outcomes. Through RBT the students can reach higher and more complex thinking skills and the concepts with which they face difficulty in learning are identified. Using this taxonomy as a framework for developing testing modules leads to more effective e-learning systems because the accomplishment of the learning objectives is better evaluated. Thus, it produces valid and efficient assessments, improving the quality of entire learning environment. Future work concerns the investigation of further adaptivity methods incorporating different artificial intelligent techniques and their evaluation.

References

1. Fletcher-Flinn, C.M., Gravatt, B.: The efficacy of computer assisted instruction (CAI): a meta-analysis. J. Educ. Comput. Res. **12**(3), 219–241 (1995)
2. Mampadi, F., Chen, S.Y., Ghinea, G., Chen, M.P.: Design of adaptive hypermedia learning systems: a cognitive style approach. Comput. Educ. **56**(4), 1003–1011 (2011)
3. Virvou, M., Alepis, E., Troussas, C.: Error diagnosis in computer-supported collaborative multiple language learning using user classification. In: JCKBSE, pp. 266–275 (2013)
4. Fallows, S., Chandramohan, B.: Multiple approaches to assessment: reflections on use of tutor, peer and self-assessment. Teach. High. Educ. **6**(2), 229–246 (2001)
5. Krathwohl, D.R.: A revision of Bloom's taxonomy: an overview. Theor. Pract. **41**(4), 212–218 (2002)
6. Hwang, W.Y., Wang, C.Y., Hwang, G.J., Huang, Y.M., Huang, S.: A web-based programming learning environment to support cognitive development. Interact. Comput. **20**(6), 524–534 (2008)
7. Kostadinova, H., Totkov, G., Indzhov, H.: Adaptive e-learning system based on accumulative digital activities in revised Bloom's taxonomy. In: Proceedings of the 13th International Conference on Computer Systems and Technologies, pp. 368–375 (2012)
8. Mayer, R.E.: A taxonomy for computer-based assessment of problem solving. Comput. Hum. Behav. **18**(6), 623–632 (2002)
9. Hubalovska, M.: Primary E-learning from the perspective of the revised Bloom's taxonomy. Int. J. Educ. Inf. Technol. **9**, 195–199 (2015)
10. Al-Rajhi, L., Salama, R., Gamalel-Din, S.: Personalized intelligent assessment model for measuring initial students abilities. In: Proceedings of the 2014 Workshop on Interaction Design in Educational Environments, p. 41 (2014)
11. Crowe, C.D., Wenderoth, M.P.: Biology in bloom: implementing Bloom's taxonomy to enhance student learning in biology. CBE-Life Sci. Educ. **7**(4), 368–381 (2008)

Author Index

M. Virvou et al. (Eds.): JCKBSE 2018, SIST 108, pp. 259–260, 2019.
https://doi.org/10.1007/978-3-319-97679-2

Printed in the United States
By Bookmasters